T0178794

BANG!!

THE COMPLETE HISTORY OF THE UNIVERSE

BRIAN MAY   PATRICK MOORE   CHRIS LINTOTT   HANNAH WAKEFORD

WELBECK

Published in 2021 by Welbeck

An Imprint of Welbeck Non-Fiction Limited,
part of Welbeck Publishing Group
20 Mortimer Street London W1T 3JW

Created by Canopus Publishing Limited, www.canopusbooks.com
Director and Editor Robin Rees
Major artworks Brian Smallwood
Diagrams James Symonds
Rear cover photo Denis Pellerin

First published by Carlton Books Ltd in 2006

A CIP catalogue record for this book is available from the
British Library

ISBN 978 1 78739 827 6

Printed in Dubai

10 9 8 7 6 5 4 3 2 1

▶ **The Heart Nebula**
Also known as IC 1805, the Heart Nebula
lies in the constellation of Cassiopeia. The
smaller nebula IC 1795. similarly ringed by
orange and yellow clouds of dust and gas,
but with a dark bar across its centre, lies at
lower right.

700 MILLION TO 9 BILLION YEA

300,000 TO 700 MILLION YEARS

10^{-43} TO 10^{-32} SECONDS

BIG BANG   TIME A.B. (AFTER THE BANG)

CONTENTS

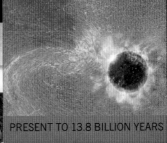

PRESENT   INFINITY?

PREFACE TO THE FIRST EDITION 2006

None of us would be here discussing the so-called 'Big Bang', had it not been for the exclamation of an astronomer who considered the whole idea to be ridiculous.

From the late 1940s, eminent British astronomer Fred Hoyle was famously an advocate of what was called the steady-state hypothesis, proposed originally by Hermann Bondi and Thomas Gold. For philosophical reasons, Hoyle was attracted to this idea that the Universe on a broad scale ought to be unchanging with the passage of time. He and others noted that the component parts of the Universe were all fleeing away from each other, a discovery made by Edwin Hubble and Georges LeMaître in the 1920s. So, to keep things looking the same, the steady-state theorists reasoned that new matter must be constantly coming into existence everywhere to replace the loss (a concept termed 'continuous creation'), with the result that the Universe would remain essentially the same forever. Meanwhile Ukrainian cosmologist George Gamow was arguing that, on the contrary, the Universe might have come into existence at a single instant, and was not in a steady state at all. In a radio broadcast in 1949, Hoyle strongly asserted that current observational evidence was in conflict with theories requiring all matter to have been created in 'ONE BIG BANG', and in doing so unwittingly coined the name that henceforth would always be used to describe the theory he spent the rest of his life fighting.

During the 50s and early 60s, a battle raged between the two theories, but gradually evidence began to pile up in favour of exactly the primeval explosion that Hoyle had found

◄◄ **Beneath southern skies**
European Southern Observatories (ESO) photographer Petr Horálek captured this photograph of the bright band of the Milky Way crossing the skies above the Paranal Observatory in Chile. The Gum Nebula (Gum 12) is the red area at top, and the Large Magellanic Cloud, a sister galaxy to the Milky Way, is the separate red patch at right, beneath the star Canopus, the brightest star in the southern hemisphere. The two figures in silhouette are ESO photographers Yuri Beletsky and Babek Tafreshi.

◄ **The original authors of Bang!**
Chris Lintott and Brian May stand behind Patrick Moore as they prepare to observe the transit of Venus in 2004.

The California Nebula
NGC 1499, the California Nebula. is located in the constellation of Perseus about 1,000 light-years away from Earth and is about 100 light-years in length. It is a classic emission nebula comprised of vast amounts of dust and gas, including hydrogen, sulphur, and oxygen, and it is the birthplace of new stars. The light emitted by the nebula is powered by Xi Persei, the bright star just to the right of the blue area of the nebula.

so unpalatable. Finally, in 1964, a mortal blow was delivered to the steady-state theory when Penzias and Wilson (unwittingly at first) discovered the Cosmic Microwave Background – the actual echo of the Big Bang itself, reverberating through the whole of Creation, billions of years later.

The Big Bang theory, (or more precisely, collection of theories), is exactly what it says; only a theory – a virtual model constructed to fit the available evidence that we find, as we observe and measure the Universe we live in. In astronomy, models come and go. The evidence is not yet all in, so we would be amazed if in a few years time our book would not need to be substantially re-written. But the story we tell in these pages constitutes what most astronomers currently believe to be a good model.

We have set ourselves the goal of telling our main story, the story of the evolution of the Universe itself, in the order in which it happened, so we decided to put the historical anecdotes and other diversions into what we have affectionately come to call 'grey areas', rather than include them in the main text. If you wish to experience the story of the Universe without interruption, feel free to skip the material in the grey areas; save them for later. Bear in mind that our main story begins in Chapter 1. Each successive chapter then describes the events of a period of time, up to the present day and beyond, into the farthest reaches of the foreseeable yet almost unimaginably distant future.

At intervals in the top right-hand corner of the pages, you will find a helpful absolute time reference – a reminder of how far we have moved along the time line. Our convention in this book is to work on an absolute timescale with zero at the moment of creation; for convenience, such times are notated as A.B. – After the Bang.

At the end of the book, there is also an introduction to practical astronomy, written by Patrick; after all, we all started by gazing up at the night sky, and wondering what it was all about.

For help and inspiration …
Our thanks to Jimmy Alvarez, Tim Benham, Sara Bricusse, Sally Avery-Frost, David Burder, Marcus Chown, Adam Corrie, Jane Fletcher, John Fletcher, Jamie Cooper, Garry Hunt, Petr Horálek, Greg Parker, Roger Prout, Phil Webb, J-P Metsävainio, Woodstock typewriters … and of course Ptolemy and Jeannie.

Note on units
In keeping with current practice, we use the term 'billion' throughout to mean one thousand million. Temperature is measured in degrees Celsius (Centigrade) or Kelvin (Celsius plus 273 degrees). To convert Celsius to Fahrenheit, multiply by 1.8 and add 32.
The unit of astronomical distance is the light-year, equivalent to about 6 thousand billion miles or 9.6 thousand billion kilometres.

PREFACE TO THE
NEW EDITION

Since *Bang!* was first published in 2006, much has changed. Shortly after the last revisions to the text in 2009, we lost Patrick – and his desire to communicate clearly and have fun while doing so are still missed every day. He would have marvelled at the discoveries of the last decade or so, ranging from a helicopter on Mars to new ways of detecting black holes deep in the cosmos, and we hope he would have approved of our attempt to bring this edition of *Bang!* up to date.

We are a team of three again, with Brian and Chris joined by Hannah Wakeford. Hannah's expertise is in the study of exoplanets – worlds in orbit around stars other than the Sun – an area of astronomy where more rapid progress has been made in the last 15 years than in any other. Remarkably, understanding these systems is leading us towards new ideas about the history of our own Solar System, and we've had a lot of fun learning and then telling this new story.

As predicted in the original preface to the first edition, progress has been rapid in many different areas. With new telescopes and technologies under development and construction, it feels as if astronomy is on the edge of another revolution. A good time, then, to take stock of what we know – and what we don't – about the Universe that surrounds us. Let's start at the beginning.

Brian, Chris and Hannah
April 2021

Check the BANGUNIVERSE website for
current activities and discussions.

INTRODUCTION **The Lure of the Skies**

Look upward on a dark, clear night and you will see stars – hundreds of them, even thousands, if you are lucky enough to live away from the light pollution of our modern cities. The heavens seem to be ablaze. We now know that these tiny, twinkling points of light are suns, many of them far larger, hotter and more powerful than our own Sun, and that our Earth appears to be an insignificant planet, perhaps less important in the wide Universe than a single grain of sand in the Sahara. But what lies behind it all? How did the Universe begin? How is it evolving, and how will it die, if indeed it ever comes to an end?

Astronomers are attempting to answer these questions, and it is truly remarkable that such minuscule beings, living on a small planet moving round an undistinguished star, have been able to peer into the depths of space, pick up light from star-systems unbelievably far away, and even send machines to other worlds. It may well be that other civilizations can outmatch ours, and that we must be regarded as cosmic primitives, but as a race we have at least made a start in coming to an understanding of the Universe we live in. In this book we are going to do our best to tell the story of the Universe – from its creation, long before the Earth existed, through to the present day, and then into the future, when the Earth will no longer be even a memory. There is much that we do not know, and perhaps will never know, but we have come a long way since our ancestors gazed up toward the stars, just as we do today, and wondered what they were.

We are living in a golden age of astronomy. Instruments such as the Hubble Space Telescope, which orbits the Earth beyond the haze of our atmosphere, would have been unimaginable only a few decades ago. Another factor which has been crucial in the amazing advances of the last fifty years is the increase in computer power available to scientists.

Across astronomy, there has been spectacular recent progress. We have sent a spacecraft past Pluto, now know of nearly 5,000 planets around other stars, and can detect ripples in space caused by the merging of distant black holes. In cosmology – the study of the past, present and future of the Universe – the old picture of a static Universe has been replaced by one which reveals a dynamic and ever-changing cosmos.

Where are we?

In our story we deal with immense distances and vast spans of time. The Earth, a globe about 8000 miles (12,800 km) in diameter, moves round the Sun at a distance of 93,000,000 miles (150,000,000 km); it is one of eight major planets which, together with many more less substantial bodies, make up the Solar System.

Most of the planets have satellites; we have one, our familiar Moon, which is our faithful companion in space, and stays with us in our constant journeying round the Sun. Like the planets, it shines only by reflected sunlight; it lies a mere quarter of a million miles (400,000 km) from us, which is why it looks so imposing. It is the only other world that has been reached by human beings, and nobody who was alive in 1969 will ever forget the sensation of triumph as Neil Armstrong took 'one small step for [a] man, one giant leap for mankind' on to the bleak rocks of the lunar Sea of Tranquillity.

But the Solar System is a very minor unit in the Universe; our Galaxy, known as the Milky Way, contains at least a hundred billion (in this book one billion equals one thousand million) suns, and we know that most of these are attended by planets. We do not know whether these planets are home to any kind of life, let alone thinking beings.

▶▶ Jets from the Necklace Nebula

The Necklace Nebula is a planetary nebula, a gas
cloud emitted by a star towards the end of its life.
The diamond-like jets forming the necklace are
knots of glowing gas. At the centre of the necklace
are two stars orbiting each other so closely they
appear as one.

▼ Orion

The brightest stars of this magnificent constellation,
always prominent in the night sky in winter,
have suggested the strong shape of a man since
prehistoric times; we know him as Orion the Hunter.
The bright orange star, at the top left of the figure
– his shoulder – is Betelgeux. The blue-white star
Rigel, at bottom right marks a leg. Midway between
them lie three stars almost in a line, known as
Orion's Belt. The smaller line of stars running from
the belt downward makes up Orion's sword, and
includes the Orion Nebula, our nearest star-forming
region, visible to the naked eye as a misty light
around the central star of the sword.

At the speed of light

The stars are remote. To try to give their distances in miles or kilometres would be as
clumsy as giving the distance between London and New York in inches, but fortunately
there is a better unit to hand. Light does not travel instantaneously; it flashes along at
the rate of 186,000 miles per second (300,000 kilometres per second), so that in one year
it covers nearly 6 thousand billion miles (9.6 thousand billion kilometres). This is what
has become known as the light-year (note that it is a unit of distance and not time). The
nearest star beyond the Sun is just over 4 light-years away, while the most remote objects
so far recorded are over 12 billion light-years from us.

Seen across such vast distances, the stars shine merely as tiny points of light.
Appearances are deceptive; many of the stars visible on any clear night are not only
much more luminous than the Sun, but also much larger. For example, Betelgeux in the
constellation of Orion, over 300 light-years away, is vast. Its globe could easily contain
the entire orbit of the Earth round the Sun. Some features have been detected on its
surface, but only the Sun is near enough to be studied in real detail – and much of our
knowledge of the stars in general depends upon what we have learned from studying our
own neighbourhood star. Fortunately, the Sun is a very normal star, neither particularly
powerful nor particularly feeble, and certainly not as variable as many. Astronomers rank
it as a dwarf, but in fact it seems to be slightly more massive than the average – and giant
stars such as Betelgeux are much less numerous than the dwarfs.

We can also gain a great deal of understanding by looking at the colours of the stars.
Just as we talk of objects being red- or white-hot, with white-hot objects being hotter

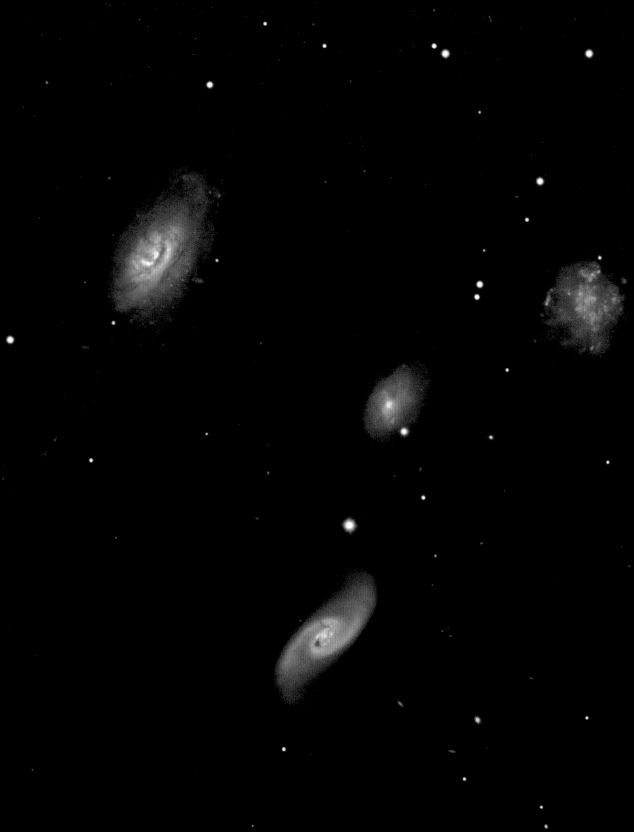

◀ **Hubble Space Telescope**
Barely skimming the Earth's atmosphere at a height of 380 miles (600 km), this unique telescope has utterly transformed our knowledge of the Universe since its launch in 1990.

than red-hot ones, so the colours of the stars reflect their temperatures. Betelgeux, for example, appears red because it is cooler than our own Sun, whereas Rigel – the other bright star in Orion – is blue-white and is much hotter than our own, yellow Sun, which is intermediate in temperature as well as in size.

The history of time

Because of the vast distances involved, when we look at the stars we are taking part in time-travel, without any need for a Wellsian machine or Dr Who's Tardis. Consider Sirius, which is the brightest star of the night sky and is very conspicuous for several months in each year. It is 26 times as powerful as the Sun, and 8.6 light-years away – that is to say, roughly 50 thousand billion miles. Its light takes 8.6 years to reach us, so that if we look at it in the year 2021 we are actually seeing it as it used to be in 2013.

The Pole Star (Polaris), which is well known to navigators, is about 430 light-years from us according to the latest measurements. The light we see coming from it right now left Polaris around 1591 – and any astronomer there equipped with a sufficiently powerful telescope could look at the Earth and see England as it used to be in the time of Shakespeare.

The light now reaching us from Rigel started on its journey towards us in the time of the Crusades, and even this is local by the standards of the Universe as a whole. We can now study objects that are so remote that we see them as they used to be long before the Earth existed.

◀◀ **Robert's Quartet**
This compact group of four galaxies is about 160 million light-years away. Apart from their beauty, such small groups are excellent laboratories to study galaxy interactions.

▶ Looking back to the Big Bang

Looking out into space, we are literally looking back in time. The light we see from the planets in our Solar System left them only minutes ago, but when we observe the most distant galaxies the Hubble Telescope has been able to photograph, we see them as they were 12 billion years ago. In this schematic diagram we, the observers, are at the bottom of the picture looking up. We will never be able to see light from the Big Bang itself, but we believe we know when it happened. On this diagram it provides the top, most-distant point – the beginning of time. The vertical scale of distance we've used here is not 'normal' or 'linear', like a tape measure. It's what's known as a 'logarithmic' scale … which means roughly that the further upwards we go, the more the distance scale is compressed, and at an ever-increasing rate. It actually doubles every few millimetres. So a vertical millimetre at the bottom represents about a thousand miles (kilometres) – but at the top it represents billions of miles (kilometres) and more.

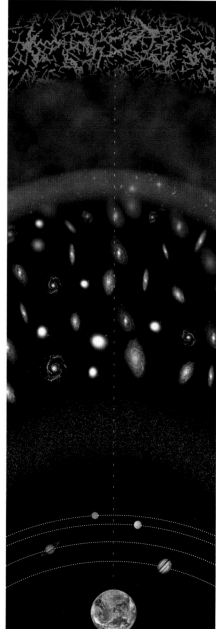

Big Bang, 13.8 billion years ago.

Opaque Universe, up to 13.4 billion years ago.

The Dark Ages.

First stars and galaxies formed 13.2 billion years ago.

Most distant observed galaxies. Light left them 12.7 billion years ago.

Many of the galaxies we see formed in this period.

Nearby galaxies, such as the Virgo cluster. Light left these galaxies 50 million years ago.

The edge of our Milky Way Galaxy.

Nearest major galaxy Andromeda is about 2.5 million light-years away.

Nearest star, Proxima Centauri, is 4.3 light-years away.

We see Jupiter as it was roughly an hour ago.

You are here.

This time travel is essential for our understanding of the Universe; we can actually see much of the story we want to understand. For example, if we suspect galaxies were much smaller in the past, we can actually observe them to confirm this. By looking at galaxies six billion light-years away, we believe we are seeing the Universe that our own galaxy inhabited six billion years ago.

If the distance scale stretches the imagination, the timescale is equally staggering. Various lines of investigation tell us that the age of the Earth is approximately 4.6 billion years and that it was formed from a cloud of dust and gas surrounding the youthful Sun, but we humans are newcomers to the terrestrial scene. To drive this home, let us imagine a timescale in which the age of the Earth is represented by one year: it came into existence at midnight on January 1. Primitive life appeared by early May, but fish did not evolve until mid-November, and the first forays on to the land surface were achieved at the very end of November. Reptiles ruled the world during the first weeks of December; the dinosaurs died out around December 15 while mammals came unobtrusively into the picture, but only on the morning of December 31 did the first, distant ancestors of modern humans arrive. The whole story of *Homo sapiens* is compressed into the last hour of the last day of the year. Jesus Christ appeared on Earth less than a minute ago.

We are reasonably confident about our distance-scale and about the age of the Earth. We have also made tremendous progress in estimating the age of the Universe as we know it; the latest value of 13.8 billion years is probably accurate to within a few per cent. However, this introduces a really major problem.

The inescapable fact is that we exist; we are made up of atoms and molecules, and this material must have been created in one way or another. Either it has always existed, or else it was produced at a definite moment in time. Neither picture is easy to accept. If the material of which we are made has always existed, we have to visualize a period of time that had no beginning. If it came into existence suddenly, 13.8 billion years ago, what happened before that? *Was* there a 'before'?

The mathematical answer is that time began with the Universe, so that there was no 'before'. This may be theoretically accurate, but it is certainly unsatisfying. In studying the Universe we treat time as a fourth coordinate – the original draft of the book was written at latitude 50 degrees north and longitude 0.41 degrees west a few metres above sea level, but in order to find the authors you also need to specify a time – late 2006.

Yet this simple picture breaks down over astronomical scales. Let us say that in the far future astronomers want to carry out an experiment simultaneously on Earth and on the nearest star, Proxima Centauri, which is located a little more than four light-years away. As no information can travel faster than light, a light signal sent between the two systems will not suffice to coordinate the experiment – time is not an absolute thing upon which all observers will agree.

In the uncertainties that we find surrounding us, we can only make intelligent guesses. This may sound haphazard, but this is essentially the scientific method. To explain an observed fact, a theory is put forward. The theory is then used to make predictions. By making new observations the predictions can be tested. If the predictions are confirmed, we have a good theory; if not, we must think again. In the following chapters, in constructing our model of the history of the Universe, we will be using the theories which have best stood up to the scrutiny of current experimental astronomy. And so to the beginning…

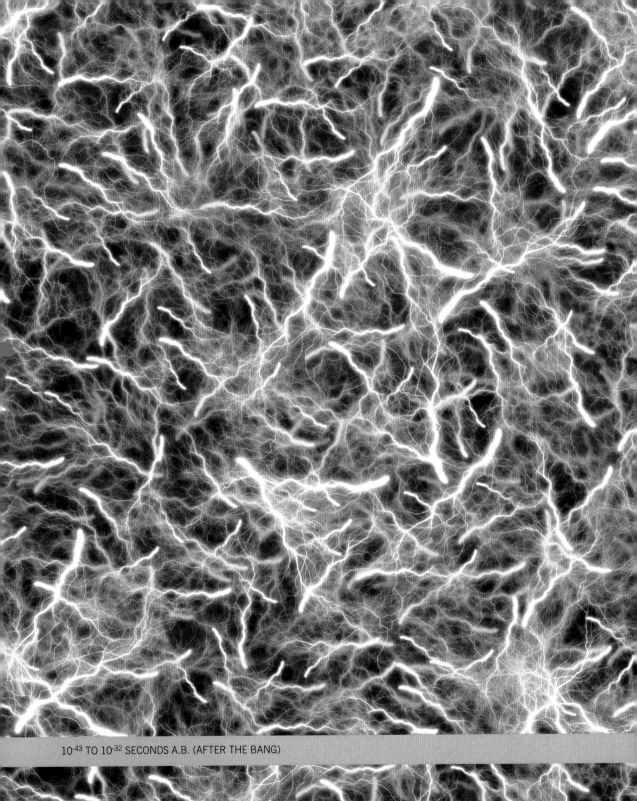

10⁻⁴³ TO 10⁻³² SECONDS A.B. (AFTER THE BANG)

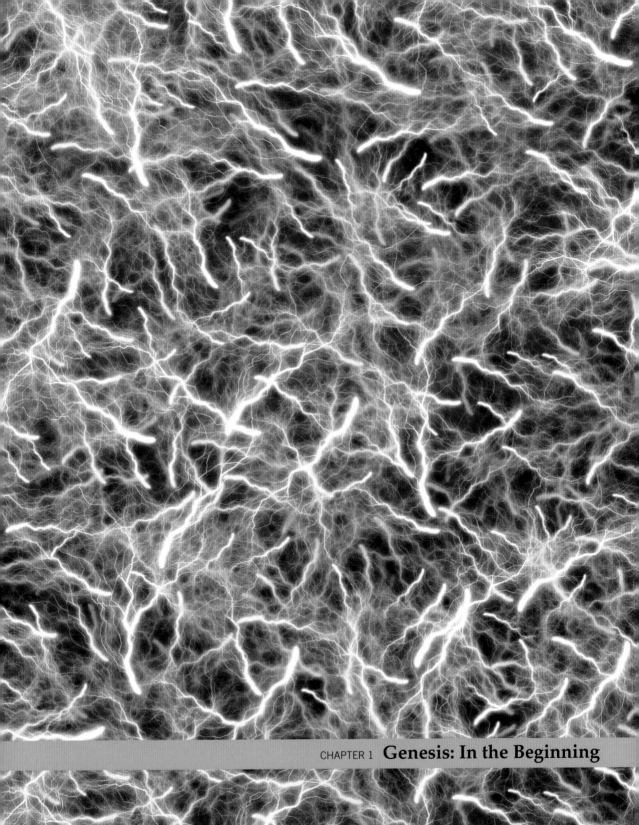

CHAPTER 1   **Genesis: In the Beginning**

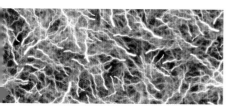

▲ Cosmic chaos

The chaotic, almost infinitely small Universe, a tiny fraction of a second after the Big Bang. In this representation the bright lines represent short-lived particles that are continually created and annihilated as they collide with each other.

▶▶ Andromeda Galaxy (M31)

Looking out through the foreground stars of our Galaxy, we see our nearest large neighbour in its entirety, comprising as many as a trillion stars. At a distance of 2.5 million light-years, it is one of the most distant objects visible to the naked human eye.

Everything, space, time and matter, came into existence with a 'Big Bang', around 13.8 billion years ago. The Universe then was a strange place – as alien as it could possibly be. There were no planets, stars or galaxies; there was only a melée of elementary particles; these filled the Universe. Moreover, the entire Universe was smaller than a pin-prick, and it was unbelievably hot. At once it began to expand – and as it spread out from this bizarre, unexpected start, it evolved into the Universe we see today.

Modern science is unable to describe or explain anything that happened in the first 10^{-43} seconds after the Big Bang. This interval, 10^{-43} seconds, is known as the Planck time, named after the German scientist Max Karl Ernst Planck. He was the first to introduce the concept that energy could be regarded not as a continuous flow, but as packets, or 'quanta', each with a specific energy. Quantum theory is now at the base of much of modern physics; it deals with the Universe on the smallest scales and is certainly one of the two great achievements of 20th-century theoretical science. The other is Einstein's general theory of relativity, which deals with the physics of very large scales – astronomical scales, in fact.

Despite the fact that, in their own realms, both theories are extremely well tested by experiments and observations, there are major problems in reconciling these theories with each other. In particular, they treat time in fundamentally different ways. Einstein's theories treat time as a coordinate; it is therefore continuous, and we move smoothly from one moment to the next. In quantum theory, however, the Planck time represents a fundamental limit – the smallest unit of time that can be said to have any meaning at all, and the smallest unit that could ever, even in theory, be measured. Even if we built the most accurate clock possible, we would see it jump rather erratically from one Planck time to the next.

Attempting to reconcile these two contrasting views of time is one of the major challenges for 21st-century physics (theories such as string theory, and its competitors, attempt to do just this, but there is as yet no version of such a theory that we can test either by experiment or observation). For now, in the small, hot, dense Universe that existed just after the Big Bang, quantum physics holds sway, and hence we begin our scientific study of the Universe 10^{-43} seconds after the beginning.

The Big Bang is a counter-intuitive idea. Our common sense seems much more impressed by a static and infinite Universe, and yet there are good scientific reasons to believe in this singular event. If we accept the Big Bang, it is possible to trace the whole sequence of events from that first Planck time right up until the present day, where we find ourselves on what Carl Sagan memorably described as our 'pale blue dot'.

The beginning of time

So let us look back to the very start of the Universe – just after the Big Bang itself. It is tempting to picture the Universe suddenly bursting out in a vast ocean of space, but this is completely misleading. The true picture of the Big Bang is one in which space, matter and,

Standard form

10^{-43} is a convenient form of mathematical shorthand, and could be written as a decimal point followed by 42 zeros and then a 1. This is precise enough, but is decidedly clumsy (0.000000000000 0000000000000000000000000000001). In order to deal with the incredibly large and incredibly small numbers that pervade astronomy, we will use so-called standard form throughout the book. 10^{33}, for example, is 1 followed by 33 zeros – we could write 1000 as 10^{3}, or 0.001 as 10^{-3}.

▶▶ Large-scale structure in the Universe

This enormous cluster of galaxies (AC03627), 250 million light-years away, is typical of what we might see in every direction, if we could see past the dust and gas of our own Galaxy and its neighbours. Clusters such as these are the largest objects in the Universe to be held together by their mutual gravitational pull.

▼ Cubic Space Division

Dutch artist M. C. Escher created this lithograph in 1952. Born in Holland in 1898, his work became internationally recognized after his first important exhibition was reviewed by *Time* magazine in 1956. Mathematicians recognize the extraordinary visualization of their abstract principles in his work.

crucially, time were born. Space did not appear out of 'nothingness'; before the moment of creation there was no 'nothingness'. Time itself had not yet begun, and so it does not even make sense to speak of a time before the Big Bang. Not even a Shakespeare or an Einstein could explain this in plain English, though the combination of the two might be useful!

It also follows that when we survey the Universe today, it is meaningless to ask just 'where' the Big Bang happened. Space only came into existence with the Big Bang itself. Hence, in those first few fractions of a second, the entire Universe we see today was in a tiny region, smaller than an atomic nucleus. The Big Bang happened 'everywhere', and there was no central point.

A nice illustration of this is given in a famous painting by Escher, known somewhat unromantically as *Cubic Space Division*. Imagine standing on any of the cubes that mark junctions on this lattice, while each and every one of the rods joining the cubes expands. From your perspective, it would seem that everything is rushing away from you, and it might seem natural at first to conclude that you are in a special location – the centre of the expansion. Yet standing back and thinking allows you to realize that the expansion would look the same wherever in the lattice you were; there is no centre. The situation is very similar in our Universe; each group of galaxies appears to be rushing away from us, and yet observers looking back at us from these distant stars would see the same illusion and would presumably be just as likely to conclude that they are all at the centre of the expansion.

Another problem concerns the often-asked, and at first glance, sensible question, 'how big is the Universe?' Here we again have a major problem – there seem to be two possible answers. Either the Universe is of finite size, or else it isn't. If finite, what lies outside it?

▼ **From the invisible to infinity**

We can now study objects at both extremes in scale. The size of human beings is between one and two x 10^0 (i.e. one) metres. If we measure the size of the Earth, it comes in at a few x 10^6 (i.e. a few million) metres. Our range of experience ranges from about 10^{-15} metres, the scale of fundamental particles that make up atoms – quarks, etc, to 10^{25} metres, the scale of the entire observable Universe.

The question is meaningless – space itself exists only within the Universe, and therefore there is literally no 'outside'. On the other hand, to say the Universe is infinite is really to say that its size is not definable. We cannot explain infinity in everyday language, and neither could Albert Einstein (we know, because Patrick asked him!).

Remember, too, that we need to consider time as a coordinate; in other words, we cannot simply ask 'how big is the Universe?' as the answer will change over time. We could ask 'how big is the Universe now?', but as we shall see later, a consequence of relativity is that it is impossible to define a single moment called 'now' that has the same meaning across the entire Universe.

Talking about a Universe that has a particular size immediately leads to thoughts of an edge. If we travelled far enough, would we hit a brick wall? The answer is no – if the Universe is what mathematicians call finite but unbounded. A useful analogy is that of an ant crawling on a ball. By travelling always in the same direction on its curved surface it will never come up against a barrier, and it can cover an infinite distance. This is despite the finite size of the ball, to which the ant will be completely oblivious. Similarly, if we were to set off in a powerful spacecraft in what we perceive to be a straight line, we would never reach the edge of the Universe – but this does not mean that the Universe is infinite; we shall see later that space, too, may be regarded as curved.

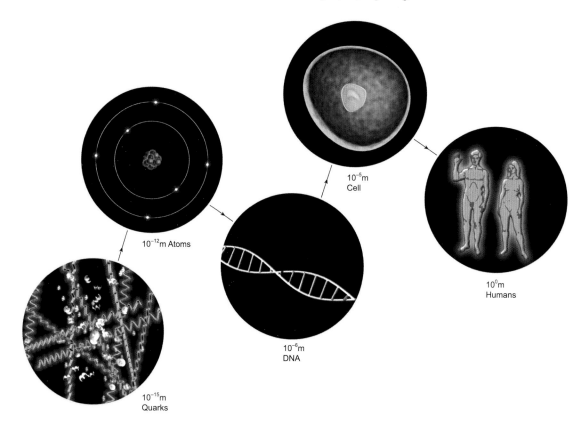

10^{-5}m
Cell

10^{-12}m Atoms

10^0m
Humans

10^{-6}m
DNA

10^{-15}m
Quarks

So let us restrict ourselves to questions we can answer scientifically, meaning questions we can answer through comparison with observation. We can say for certain that the *observable* Universe (literally that part of the Universe from which light can potentially have reached us) is finite in size because, at our current best guess, the Universe is only 13.8 billion years old. Therefore the edge of the observable Universe, from where light could only just be reaching us, must be 13.8 billion light-years away, and expanding at a rate of one light-year per year. In fact there are reasons why we will never be able to see quite this far, as will become clear later. All we can say for certain about the size of the Universe is that it must be larger than the portion we can see.

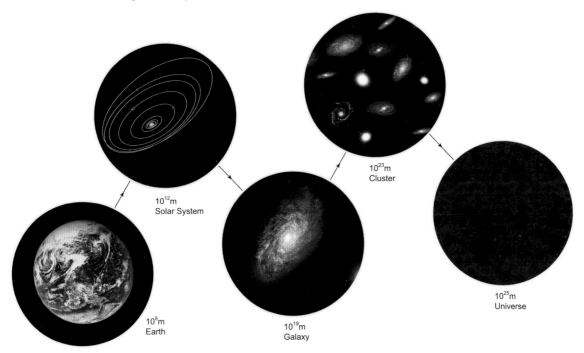

10^{12}m
Solar System

10^{23}m
Cluster

10^6m
Earth

10^{19}m
Galaxy

10^{25}m
Universe

The scale of the Universe

Of course, saying that an object is 13.8 billion light-years away is all very well, but can we really comprehend the scale of the Universe? It is possible to appreciate fully the distance between, say, London and New York, or even the distance between the Earth and the Moon – roughly a quarter of a million miles – which is roughly ten times the circumference of the Earth, and many people have flown for a greater distance than that over their lives. Indeed, several airlines give special privileges to those who have travelled more than a million miles in their lifetimes. But how do you really grasp 93 million miles, the distance of the Sun? And when we consider the nearest star, 4.2 light-years (approximately 24 million million, or trillion, miles) away, we are quite unequal to the

▲ **Atomic layers**

A microscope image of atomic layers on the surface of a crystal of iron silicide. The smallest step is only one atom thick.

task…. The galaxies are immensely more distant than this: even the Milky Way's nearest neighbours, such as the Andromeda Galaxy, are over two million light-years away.

At the other end of the scale, visualizing the size of an atom, which cannot be seen individually with any ordinary microscope, is equally difficult. It has been said that, in scale, a human being is about half-way between an atom and a star. Interestingly, this is also the regime in which physics becomes most complicated; on the atomic scale, we have quantum physics, on the large scale, relativity. It is in between these extremes where our lack of understanding of how to combine these theories really becomes apparent. The Oxford scientist Roger Penrose has written convincingly of his belief that whatever it is that we are missing from our understanding of fundamental physics is also missing from our understanding of consciousness. These ideas are important when one considers what have become known as anthropic points of view, best summarized as the belief that the Universe must be the way it is in order to allow us to be here to observe it.

Another useful question to ask is how many atoms there are in the observable Universe. One estimate has come up with a total number as large as 10^{79}, or in other words a 1 followed by 79 zeros.

Traditionally we have viewed atoms as being made up of three more fundamental particles: the proton (carrying unit positive electric charge), the neutron (no charge at all) and the much less massive electron (carrying unit negative charge). Incidentally, it is far from easy to define what electric charge is on the atomic level. It will suffice to think of charge as a property that particles can have, just as they have a size and a mass. Charge always comes in parcels of fixed size which we call unit charge.

Classically these particles are considered to be arranged like a miniature solar system, with electrons orbiting a central compound nucleus, containing protons and neutrons. This nucleus carries a positive charge, which is exactly balanced by the combined charge of the orbiting electrons. In our Solar System of planets, the force of gravity keeps the planets in their orbits around the central Sun, but here in the atom it is the attraction between the negatively charged electron and positive nucleus that keeps the electrons in their orbits.

In passing, we should note that this simple picture can explain much of basic chemistry; for example, why it is the outer electrons of atoms that tend to be involved in chemical reactions. They are further away from the nucleus, and therefore they are less tightly held by its attractive force. So the simplest atom, that of hydrogen, has a single proton as its nucleus, and one orbiting electron. The whole atom is therefore electrically neutral: plus one added to minus one equals zero. All atoms have an equal number of electrons and protons. Each element has a unique number of these particles, known as the atomic number. For example, helium atoms have two protons and electrons – an atomic number of two – while carbon atoms have an atomic number of six. Heavy elements have large numbers of protons and electrons. Uranium, the heaviest natural element on Earth, has an atomic number of 92.

This view of the atom, which saw protons and neutrons as solid lumps, prevailed in the early 20th century, but things are much less clear-cut today. Much of the strange behaviour of extremely small systems can only now be explained by considering them to be made up of waves rather than particles. This theory is known as wave-particle duality. In addition, experiments have shown that while electrons seem to be truly unbreakable, protons and neutrons are not in fact fundamental – they can be split into smaller particles known as quarks, which themselves are now believed to be fundamental. No one has ever seen a quark, but we know they must exist since they have been detected in particle accelerators that have been built to smash protons together at incredibly high speeds. In these experiments protons are seen to fracture, and hence scientists conclude that they cannot be fundamental. Nature abhors a naked quark; quarks appear only in pairs or triplets.

The forces of nature

The reason for this property of quarks lies with an unusual property of the force that normally binds quarks together, known (not without reason) as the strong nuclear force. It is dominant on very small scales, which is why we need such powerful particle accelerators to smash protons apart. Unlike the forces with which we are familiar on larger scales, such as gravity or the attraction between opposite electric charges, the strong force increases with distance. In other words, if we could separate two quarks we would find them being pulled together more and more strongly as the distance between them increases. Eventually, as quarks move apart from each other the energy caught up in straining to pull them apart becomes so great that two extra quarks are produced, the energy being converted into mass. Suddenly we have two pairs of quarks rather than the individual quark we were attempting to isolate. This process means that no experiment ever produces individual quarks, and in the everyday Universe they exist only as components of other particles, such as protons and neutrons, which each contain three quarks.

At the huge temperatures in the Universe immediately after the Big Bang, the quarks had enough energy to roam free, and so by understanding the story of the Universe on the

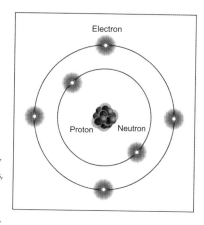

▲ **The classical atom**

This simple model of an atom was proposed by Niels Bohr in 1913. It involves a nucleus made up of neutrons (blue) and protons (red) that is surrounded by electrons orbiting like the planets around the Sun. Despite the advent of quantum mechanics, which paints a very different picture of the atom and shows fuzzy 'probability densities' in place of the ball-like particles, Bohr's classical model is still useful today.

▶ **Hunting quarks**

▶ **Hunting quarks**

At Brookhaven National Laboratory in New York, beams of gold nuclei are smashed together at close to the speed of light. The result is a recreation of a state of matter that is believed to have existed ten millionths of a second after the Big Bang, known as a 'quark-gluon plasma'. This image, looking remarkably like a human eye, shows the tracks of about 1000 particles emerging from a collision. It is effectively a cross-section; the particles are actually sent off in all directions.

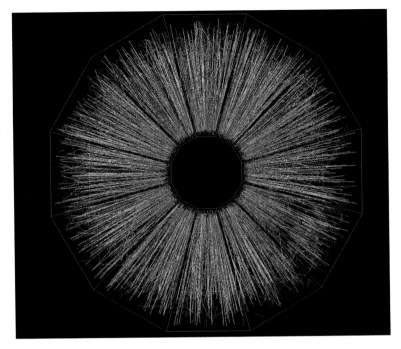

largest scales we may come to understand more about the particles that account for the smallest scales. The energy that each particle possessed in the early Universe will remain far beyond the reach of our particle accelerators; even an accelerator the size of the Solar System would be incapable of producing particles with this enormous energy.

It is a remarkable fact that our present day research into the very small, via particle physics, and into the very largest scales, via cosmology, are so intertwined. To understand the whole Universe, we are dependent on our understanding of the fundamental particles, and the best tests for our theories about them are in the embryonic Universe. A 'hot space' full of these highly energetic fundamental particles is the earliest image we can conjure of our newly-born Universe.

Bigger is cooler

From the first Planck time onwards, this inconceivably small, inconceivably hot Universe began to expand and hence also to cool. The Universe was a sizzling ocean of quarks, each of which had a vast amount of energy, moving at a huge speed. As a result, there could be no atoms or molecules of the kind we know today, because these are complicated structures, quite unable to survive the disruption of very high temperatures; the quarks were simply too energetic to be captured and confined within protons or neutrons. Instead, they were free to career around in the infant Universe, until they collided with their neighbours. As well as quarks, this early soup of sub-atomic particles also contained antiquarks – identical twins, but bearing opposite electric charges. It is now believed that each particle has its equivalent antiparticle, identical in all respects other than its electric charge, which is opposite. The

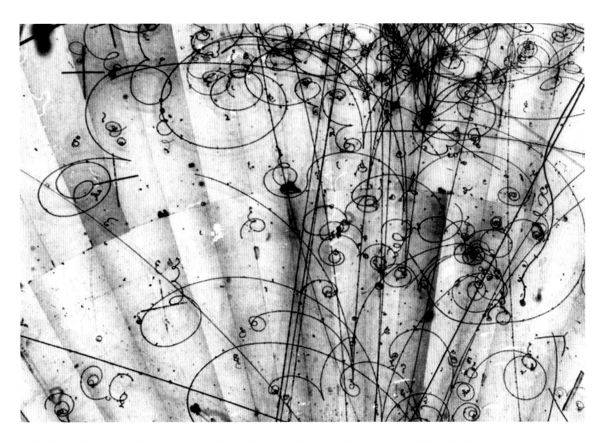

antimatter particle corresponding to an electron is a positron, which bears a positive charge, but is otherwise exactly similar to the electron. The concept of antimatter is familiar from science fiction, where it forms the basis of countless highly-advanced starship engines, all of which are based on the fact that a collision between a particle and antiparticle results in the annihilation of both and the release of much energy – this has been verified by experiments. Whenever a quark met an antiquark in the primeval Universe both would vanish, releasing a flash of radiation. The reverse process also occurred; radiation of sufficiently high energy (certainly at the energies found at this early stage of the evolution of the Universe) could spontaneously produce pairs of particles, each pair composed of a particle and its antiparticle. The Universe at this epoch, then, was composed entirely of radiation that produced pairs of particles, which in turn vanished as they collided with each other, returning their energy to the background radiation.

As the Universe continued to expand and cool, after the first microsecond (only ten million million million million million million Planck times), when the temperature dropped below a critical value of about ten million million degrees, the quarks slowed down enough to enable them to be captured by their mutual (strong force) attraction. Bunches, each of three quarks, clumped together to form our familiar protons and

▲ Seeing antimatter

This image shows the creation of pairs of electrons and positrons in a bubble chamber. Charged particles leave a trail of minute bubbles behind them, allowing the eye – or the camera – to track their progress. The photons that provide the energy to create the electrons and positrons cannot be seen, as they have no charge. Each pair of trails begins at a common point, from which the particles can be seen to spiral outwards. Moving in a strong magnetic field, the electrons and positrons experience opposite forces, and so spiral in opposite directions.

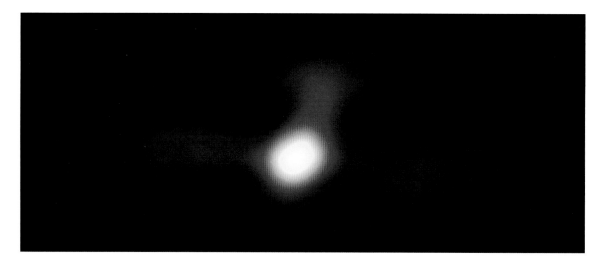

▲ Cloud of antimatter

This gamma-ray map of the centre of our Galaxy is believed to show matter and antimatter particles – electrons and positrons – colliding with each other, producing the annihilation of both sets of particles, and the emission of huge amounts of energy. This implies that antimatter in the form of positrons is streaming from the centre of our Galaxy.

neutrons (collectively known as baryons), whereas the antiquarks clumped together to form antiprotons and antineutrons (antibaryons). Had the number of baryons and antibaryons been equal, the most likely outcome is that collisions between them would have resulted in complete annihilation. The energy left in the radiation would have become diluted as the Universe expanded, and so new particle pairs would no longer have been created. Matter in the Universe would not have survived to the present day.

It was only the fact that there was a very slight imbalance built in from the very start that saved matter and therefore enables us to be here, wondering what happened in these remote times. Due to reasons we do not yet understand, for every billion antibaryons there were a billion *and one* baryons, so that when the grand shoot-out was over, almost all the antibaryons had vanished, leaving behind the residue of protons and neutrons which make up the atomic nuclei of today.

The cosmic conspiracy

Let us return briefly to the present; consider two galaxies, each nine billion light-years away from us, but in opposite directions as seen from the Earth. The distance between them is therefore 18 billion light-years. Both will exist in regions of the Universe that, broadly speaking, look the same on the largest scales. One may be deep in the heart of a cluster of galaxies, such as our own nearby Virgo Cluster, whereas the other may be much more isolated. And yet near the first there will be isolated galaxies, and near the second galaxy there will inevitably be a galaxy cluster. Each half of the Universe will contain the same types of galaxies in the same proportion, and even the temperature of the local region will be the same in both cases.

The Universe is less than 18 billion years old (remember the best estimate is 13.8 thousand million years), so light would not yet have had enough time to make the crossing between the two galaxies, and relativity insists that light is the fastest thing in the Universe. If light has not yet had time to cross the space between the two regions, then nothing else could have done, and so nothing could have passed from the first region

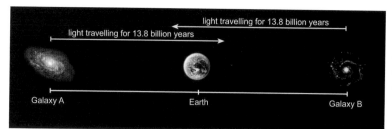

light travelling for 13.8 billion years

light travelling for 13.8 billion years

Galaxy A

Earth

Galaxy B

◀ **The cosmic conspiracy**

While we can see remote galaxies A and B in opposite directions in the sky, they cannot see each other. In the whole of the time since the Big Bang, light has still not had time to travel between them.

to the second. Any differences between the regions could not be ironed out, and it is therefore surprising that the Universe seems much the same in whichever direction we look; we see galaxies of the same type, distributed in the same way. This fact is a problem known as the 'cosmic conspiracy'.

Why is this a problem? Doesn't it seem natural that the Universe appears the same in whichever direction we look? Perhaps there is some as-yet-unknown law, governing the physics of the Big Bang itself, which ensures that only universes that are almost completely uniform can be produced. However, we have no hint of any physics that could predict this, and so we must at least consider the possibility that the Universe began with large differences in temperature between different regions, for example, an early Universe in which one half is twice as hot as the other half. How could this lead to the kind of uniformity we see today? There has not been time for heat to flow from the warm to the cool region of the Universe, and there has not even been time to send a message, travelling at the speed of light. In such circumstances, correcting this original imbalance would seem to be impossible and yet these widely separated, disconnected areas are, in fact, similar.

Our two galaxies may be far apart now, but when the Universe was very young it was also much smaller and bodies on opposite sides could have been in touch and able to exchange heat, producing the uniformity seen today. The question now, therefore, is how big the Universe was in these early stages. Surprisingly this seems to be a fairly simple question to answer.

Only one of the forces that we have so far discussed can act over astronomical distances and that is gravity, which by its very nature is an attractive force, pulling material together. Gravity alone would slow down the expansion from its initial rate. We can attempt to work backwards from the present day to determine how the size of the Universe has changed with time – and we discover that the cosmic conspiracy survives into the early Universe. In other words, the Universe was never small enough to allow light to cross from one side to the other, and therefore *never* small enough to allow temperature differences to even out. This whole scenario presupposes that gravity is the only force affecting the rate of expansion, and it is only if we are prepared to abandon this idea that we can solve the conspiracy problem.

Inflation: an extravagant solution

The currently popular solution involves increasing the complexity of the Big Bang theory somewhat. Most cosmologists now believe that there was an extremely short period of rapid expansion, known as inflation, between 10^{-35} and 10^{-32} seconds after the Big Bang, during which period the size of the Universe increased many billion times. At the end of

▲ The Hubble Extreme Deep Field

The Hubble Extreme Deep Field studied galaxies
seen at a distance of 13 billion light-years. The
distance between the most remote galaxies in the
southern and northern skies is so great that light
would not have had time to get much beyond
halfway between them, and yet the galaxies look
similar. A striking example of the cosmic conspiracy.

the inflationary period, the expansion settled back to a relatively stable rate, consistent
with that observed today.

Without a period of inflation, regions of the Universe we see on opposite sides of the
sky would neither have had time to exchange heat nor to settle down into a comfortable
equilibrium. The suggested rapid expansion allows us to believe that the Universe
was initially much smaller, and could therefore reach a uniform temperature before
the acceleration began. Any remaining small differences would then be ironed out by
the vast increase in scale. This is because another consequence of the amazingly rapid
inflation is that the region of the Universe we observe is only a tiny fraction of the entire
Universe. In other words, we are only looking at variations in what is effectively our local
neighbourhood and these are bound to be small. To give an analogy much nearer home,
the Earth as we see it is marked by huge variations in height; from the peak of Mount
Everest to the bottom of the deepest ocean trench. Inflation achieves the equivalent effect
of blowing up the patch of ground under your little toe to the size of the entire globe (or
equivalently, shrinking us to a size much smaller than the smallest virus). The differences
in height that we can reach and explore are then bound to be slight; inflation has exactly
the same influence on temperature fluctuations in the Universe.

But why should the infant Universe suddenly undergo such an extreme increase in the
speed of its expansion? It seems that there is a need to introduce a new force, which can
be held responsible for the vast acceleration, acting in the opposite direction to gravity.
Scientists have begun to study in detail what properties such a force might have and
yet there seems to be no obvious explanation. As far as we know, there are no particular

circumstances peculiar to the epoch just prior to inflation, and the appearance and sudden disappearance of this accelerating force therefore seems to be somewhat arbitrary, but it does at least allow us to deal with the problem of the 'cosmic conspiracy'.

Are there other problems which the introduction of inflation can solve for us? It turns out that inflation can also explain two other features of the Universe we see today, which are otherwise completely inexplicable. First, according to the standard theory of particle physics, a certain type of particle, known as a monopole, should occasionally appear in detectors. In fact, none has ever been detected, and this requires some explanation. The theory of inflation allows us to argue that the concentration of these particles has become so low that our failure to locate any is not surprising. Say, for the sake of argument, that 100 million million of these particles were created in the Big Bang; it would be surprising that we have failed to detect a single one. But if the same number of monopoles were created and spread through a universe that has become many thousands of millions of times larger than before inflation, it is likely that there could well be none within the entire visible Universe. The speed of inflation was so overwhelming that, even in the short time for which it operated, it produced a universe which was inconceivably larger than that predicted by the conventional Big Bang. Inflation provides an explanation for the missing particles – they have simply been diluted away.

Life in a flat Universe

There is a third pillar supporting the seemingly crazy edifice of inflation, perhaps the most convincing of all. This involves the geometry of the Universe. Most people are familiar with the geometry of Euclid which we learn, sometimes perhaps reluctantly, at school. We are told the three angles of a triangle add up to 180 degrees. However, this is not always the case; on the surface of a sphere, for example, they can add up to more than

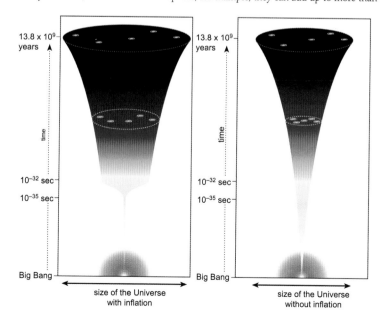

◀ **Inflation**

The left-hand diagram shows the impact of inflation compared with the Universe that did not undergo inflation (right). As the Universe expands, the galaxies move further apart. With inflation, the Universe is smaller just after the Big Bang, but much larger today.

13.8×10^9 years

13.8×10^9 years

time

time

10^{-32} sec

10^{-35} sec

10^{-32} sec

10^{-35} sec

Big Bang

Big Bang

size of the Universe
with inflation

size of the Universe
without inflation

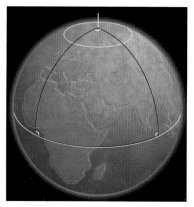

▲ Spherical geometry

In the Euclidean geometry we learn in school, the interior angles of a triangle always add up to 180 degrees. But a triangle drawn on a sphere can contain more than 180 degrees. This triangle drawn on the Earth's surface contains three right angles, adding up to 270 degrees.

180 degrees. Consider a line drawn from the north pole, down the Greenwich meridian to the equator, and then east along the equator, a turn of 90 degrees. If we complete the triangle by returning to the pole along the meridian which runs through Russia, we will make another 90 degree turn. 90 + 90 =180, and we still have to add in the angle at the top, between the two meridians. Euclidian geometry applies only to flat surfaces.

What form does the geometry of the Universe take? Things are complicated because we are dealing with a four-dimensional space (the three familiar spatial dimensions, plus time) rather than a two-dimensional surface as before. Let us consider the largest scales, and ignore local distortions caused by matter. There are huge numbers of possible geometries for the Universe, and yet it seems that our Universe has been finely tuned to select one specific type; observations show (see Cosmic Microwave Background radiation in Chapter 2) that we live in what is known as a 'flat' universe, one in which Euclidian geometry holds even on the largest scales. Why should this be so? To achieve a flat universe, we need to have exactly the correct amount of matter within the Universe, to within a few atoms. In other words, if there were a few atoms too many or too few we would be in a universe with a geometry far from flat.

Once again, we are confronted with an observation that might be ascribed to some feature of the early physics governing the Big Bang itself – and once again inflation allows us an alternative and more satisfying explanation. The argument rests, as before, on the fact that inflation provides a much larger universe than the simple Big Bang.

Let us consider a three-dimensional analogy to help us think in four dimensions. Anyone standing on top of a bowling ball will quickly discover that it is a sphere, presumably when he or she falls off. What about a larger sphere, such as the Earth, upon which we happily stand every day? It may not be immediately obvious that we are standing on a curved surface, although it is certainly easy enough to work out. Contrary to popular belief, as long ago as the time of the ancient Greeks it was known that the Earth is a sphere (they even succeeded in measuring its diameter) and just watching a ship disappear over the horizon provides a clue that the Earth's surface is curved. Now imagine being on the surface of a sphere several trillion times larger than the Earth. All our experiments would indicate that we really were on a flat surface; the curvature due to the sphere would simply be too small to be detected – our ship would take an impossibly long time to reach the horizon.

Temperature

In everyday life we mostly measure temperature in degrees Celsius (Centigrade), whose scale is based around the freezing and boiling points of water. But temperature in the scientific sense is defined differently. It is based upon the speeds at which the atoms and molecules are moving around – the faster the speeds, the higher the temperature.

Holding a thermometer in liquid, what one is really measuring is how hard the molecules of the liquid are smashing into the thermometer, and hence how fast they are moving. This definition of temperature leads to some counterintuitive results:

consider a firework sparkler and a glowing poker. Each spark is white hot, but contains so little mass that it is quite safe to hold the handle of the firework – but most people would be very reluctant to grasp a red-hot poker, even though it is much cooler than the sparks of the firework.

Another, beautiful example is the pearly white corona, the Sun's outer atmosphere, which becomes visible during a total eclipse. The visible surface of the Sun is at a temperature of around 6000 °C, while the corona (for reasons not well understood) is at around a million degrees. Yet it is

After inflation

A Universe that has undergone inflation is like this last sphere. Because it is inflated to such an immense size, our observable Universe is only a tiny proportion of it and we can measure only its local properties. We may well conclude, rightly, that the Universe we can see has a flat geometry. In such a vast universe we can know nothing about the geometry of the Universe beyond the range of our observations. Regardless of which of the many possible geometries the Universe has, inflation tells us why our measurements indicate it is flat.

These three problems are all neatly solved by the idea of inflation, although at the price of introducing this mysterious, temporary acceleration, which remains poorly understood. Perhaps when we come to a greater understanding of the Big Bang itself then we will have an alternative answer, but for now inflation seems to be as good an explanation as any.

After the end of inflation, the Universe continued expanding and cooling at a lesser rate. Around three seconds after the Big Bang, the temperature had dropped to about a billion Kelvin. About three-quarters of the material in the Universe was hydrogen, and almost all the rest was helium (remember that the helium atom has two electrons, orbiting a nucleus composed of two protons and two neutrons).

The Big Bang predicts that for every ten protons, or hydrogen nuclei, produced there was one helium nucleus. Today hydrogen atoms still outnumber helium by ten to one. This provides perhaps the most simple and yet powerful test of the Big Bang theory. Stars convert hydrogen into helium, and so we would only expect the ratio to shift in favour of helium. If we observed a single object, anywhere in the Universe, in which there was less helium than expected we would have to completely reconsider our theory. No such observation has ever been made.

Evidence for the Big Bang comes from many different directions, and is enough to convince most scientists, for now, that our Universe did indeed begin in a hot, dense state before expanding. The classic Big Bang theory does not attempt to explain why this happened – and it may be that a new theory wrought sometime in the future will be needed to explain this. For now, we must try to wrap our heads around the fact that it seems our Universe had a beginning.

▼ **Total solar eclipse – double exposure**

A total eclipse of the Sun is the grandest spectacle nature has to offer, and because of the huge range of brightness present at any one instant, no single photograph has the dynamic range to do it justice. Brian took this image of his first total eclipse in 1991 at Cabo, St Lucas, in Baja California, Mexico. By chance the filter used for the total phase was tilted at an angle to the camera lens, and by re-reflection produced a ghost image, clearly showing two prominences, which were massively overexposed in the main image – an exposure well-judged to show the beautiful shape of the fainter outer corona.

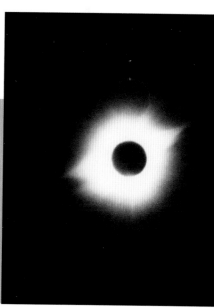

so tenuous that it would be perfectly possible to fly a spacecraft through the corona and survive. There would simply not be enough hot matter to significantly heat the spaceship.

The slower the movements of the atoms, the lower the temperature. Go down to around –273 °C (–472 °F) and the movements would have stopped completely. The temperature cannot fall further; we have reached absolute zero, the coolest possible temperature. This has never been (and never will be) attained in the laboratory, though we have worked down to a tiny fraction of a degree above

it, where matter takes on some extremely strange properties.

The Kelvin scale, named after Lord Kelvin, begins at absolute zero (0 K), but its degrees are the same size as those of the Celsius scale. To convert Kelvin to Celsius, subtract 273; to convert Celsius to Kelvin add 273. So 3 K is equal to –270 °C.

The advantage of the Kelvin scale is that we do not have to deal with negative numbers, and its zero point remains fixed and does not depend on pressure, unlike the boiling point and melting point of water.

300,000 TO 700 MILLION YEARS A.B. (AFTER THE BANG)

CHAPTER 2 **And Then There Was Light**

For the next 400,000 years following the cataclysmic period of inflation there were no major developments. The physical conditions that controlled the evolution of the Universe remained more or less constant. The Universe became a less violent place. As the temperature dropped, so the protons and neutrons began to slow down; however, radiation and matter were still linked, as we shall see. From our point of view, the biggest difference between this Universe and the Universe we see today is that in those very early times it was completely opaque.

Electromagnetic waves, including visible light, may also be regarded as a stream of photons, which are particles with zero mass that always move at 186,000 miles (300,000 km) per second. In the strange world of quantum mechanics (which is, perhaps, the best tested theory of modern science) we no longer have a clear distinction between 'waves' and 'particles', but have to accept that everything exists as something called a 'wave-particle duality', intermediate between the two. Just like the entities we traditionally think of as particles, such as electrons and protons, light behaves sometimes as a particle, the photon, and at other times as if it were a wave.

Each photon carries a well-defined 'quantum' of energy, the amount of energy being determined by the colour of the light, so that it is quite in order to say that electromagnetic radiation is 'a stream of photons'. Let us now follow the path of one of these photons, perhaps released by collision between a proton and an anti-proton in the very early Universe. In such crowded conditions, no photon could travel very far before hitting and being absorbed by an electron, which thus would gain energy. Eventually the photon might be re-emitted, but in almost all cases in a different direction from its original heading. This process would be repeated again and again, leaving the photons effectively getting nowhere very fast.

However, when the Universe had cooled to a mere 3,000 degrees, around 400,000 years after the Big Bang, a sudden change took place. Before this critical moment, the electrons – the lightest, and therefore fastest of the constituent particles of ordinary atomic matter – had been moving much too fast to be captured by the heavier atomic nuclei, but at a temperature of 3000 degrees they could no longer avoid capture. The first neutral atoms were formed. Seen on the scale of the atom, the captured electrons orbit a long way from the nucleus (atoms are, after all, mostly empty space), but compared with the distance between atoms they are very close to their nuclei. A large expanse of space between each newly formed atom therefore opened up, and photons were suddenly free to travel for great distances. In other words, matter and radiation were separated, and 400,000 years after the Big Bang the Universe became transparent.

Echoes of the Big Bang

The capturing of the electrons was amazingly sensitive to the temperature of the Universe; as soon as this dropped below the critical value, then the process occurred with remarkable rapidity. Along with the fact that the temperature of the Universe is almost exactly the same throughout the entire extent of space (thanks, remember, to inflation)

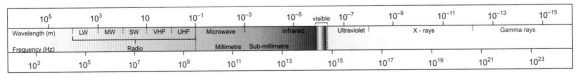

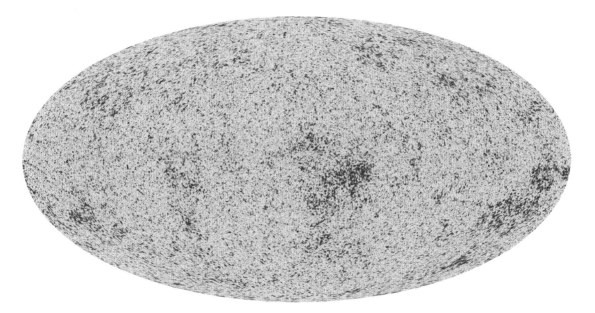

this means that the process occurred almost instantaneously across the whole Universe. The result was that light could travel uninterrupted across the Universe so that over 13 billion years later we can still see a snapshot of this particular moment in the evolution of our Universe. This ability to look back to one particular instant of time is unique in astronomy. Usually, when we try to look at the distant parts of the Universe, our view is obstructed by images of nearby galaxies, that emitted their light more recently. This magical event when the universe became transparent is observable now, without obstruction, as what we call the Cosmic Microwave Background (or CMB).

Many readers will have observed these faint echoes of the death of the 'fireball' that was born with the Big Bang, whether consciously or not. By unplugging the aerial from a television or retuning it away from a channel, you will see black and white static. One per cent of this static comes from the CMB – just under 14 billion years after being emitted it is still able to interfere with your viewing of television.

When seen today, the frequency of this background radiation is consistent with an emitter at an average temperature of only 2.7 K above absolute zero. Why so cool, if this radiation is really the echo of the Big Bang itself? The reasoning is quite straightforward; the radiation would have been emitted when the Universe was at a temperature of 3,000 degrees. As it travelled towards us, the space through which it was moving was continually expanding, stretching the light to longer and longer wavelengths, and hence leading to cooler and cooler apparent temperatures. This is our first encounter with the phenomenon known as redshift, which has come to be of fundamental importance.

The discovery of the Cosmic Microwave Background gave strong support to several predictions of the Big Bang theory. For instance, it was found that the radiation emitted conformed to that predicted for a black body, a hypothetical object which absorbs all the radiation that falls on it. If heated, it emits radiation in a spectrum in which the intensity of light at any particular wavelength depends only on its temperature. In practice, this

▲ The microwave sky

This all-sky picture of the sky observed at microwave frequencies reveals tiny temperature fluctuations when the Universe was just 380,000 years old, seen as colour differences, that correspond to the seeds that grew to become the stars and galaxies of today. Red/brown indicating warmer and blue cooler regions. The image was based on data obtained by the Planck satellite 2009–13.

▼ Big Horn antenna

The telescope with which Robert Penzias and Arno Wilson first detected the cosmic microwave background in 1964 is more accurately described as a microwave horn antenna. It is still on show at Bell Laboratories in New Jersey (without the pigeon droppings that initially confused the astronomers).

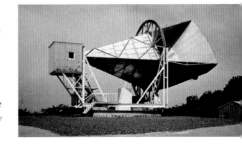

▶ The infrared sky

The top panel shows a long exposure image taken in the infrared with the Spitzer Space Telescope. At the bottom is the residual light after the subtraction of all identified foreground sources. It has been claimed that the remaining glow contains ultraviolet light emitted by the first stars, now shifted by cosmic expansion into the infrared part of the spectrum. If the claim is confirmed, this will become one of the iconic images of astronomy.

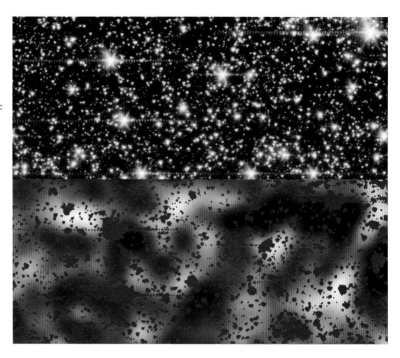

tells us something about the nature of the emitter – for example, the object would have to be isolated from external influences. The hot, dense and practically opaque Universe of the period between the Big Bang and transparency 400,000 years later would be just such an emitter. The agreement between theory and observation is now so perfect that on most plots of the data the thickness of the line used to show the prediction is larger than the uncertainty in the measurements, a situation very rare in science and unique in observational astronomy.

At first, the radiation appeared to be absolutely uniform; there seemed to be no variations linked with direction. Even after subtracting the foreground glow of microwaves emitted by our own galaxy, one part of the sky glowing in the CMB looked much the same as any other part. But the Universe we see today is 'lumpy'; there are huge distances

▶ Redshift

To understand redshift, it's convenient to regard light as a wave rather than a stream of photons. When this idea was first proposed there was a great deal of controversy – if light is a wave, what does it move in? After all, sound waves depend on air to be transmitted, and water waves can hardly exist on their own. Many people believed in a fundamental substance called the ether, which was all-pervading and within which all light travelled, but during the late 19th and early 20th centuries this was firmly replaced by the realization that light could be self-propagating and would have no need of a surrounding medium. If light is a wave, therefore, it has a wavelength, which determines both its colour and its energy. Red light, for example, is of a longer wavelength and a lower energy than green light, which itself is of a longer wavelength and a lower energy than blue light. Infrared light is radiation

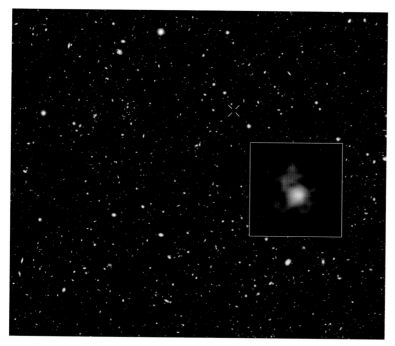

This Hubble image of the young galaxy GN-z11 shows it as it existed 13.4 billion years in the past, just 400 million years after the Big Bang, when the Universe was only three per cent of its present age, making this one of the most remote galaxies ever observed. Many more galaxies in the early Universe are expected to be found by the upcoming James Webb Telescope.

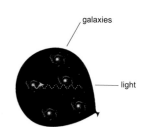

1 000 000 000 years ago

between the relatively dense galaxies, which are themselves grouped into clusters, and the clusters into superclusters. These superclusters are themselves separated by enormous voids, now beginning to be seen in detail in surveys such as the Anglo–Australian two degree field (2dF survey) and the Sloan Deep Sky Survey, which reach out a billion light-years from Earth. Whichever way we paint the picture of our Universe emerging from these observations, it is certainly not uniform, and so there was clearly something wrong. Hidden somewhere in the seemingly uniform early Universe there must be the seeds of the structure we see today.

The cosmic background radiation is now the most studied phenomenon of astrophysics, and it has much still to tell us. It marks our earliest view of structure in the Universe. The detailed view of the CMB provided by satellites like ESA's Planck revealed temperature

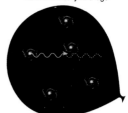

500 000 000 years ago

with a wavelength longer than that of the red light we can see, and radio waves are of still longer wavelengths. At the short-wavelength end we come to ultraviolet light, and then to X-rays and gamma rays. Since the Cosmic Microwave Background was emitted, the light we detect today has been travelling toward us through an expanding Universe. This expansion should not be thought of as objects rushing away from each other but as an expansion of space itself. As space expands, it stretches the light travelling through it, increasing its wavelength. Blue light becomes green, then red, then infrared, and we say it has been redshifted. This process can be visualized as a balloon being inflated (right). Everything on the surface moves further away from everything else. Hence the CMB, emitted in much more energetic regions of the spectrum is now detected primarily as low-energy microwaves.

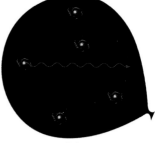

today

▶ **Traces of the Big Bang**

Captured by the Cosmic Background Explorer
satellite, (COBE) these maps show minute
temperature differences across the sky, reflecting
disconformities in the early Universe. At the top is
mapped the raw data, the middle map has the effect
of the Earth's motion through space removed, and at
the bottom we see the result of compensating also
for radiation from the Milky Way, leaving only the
temperature differences resulting from the remains
of the Big Bang.

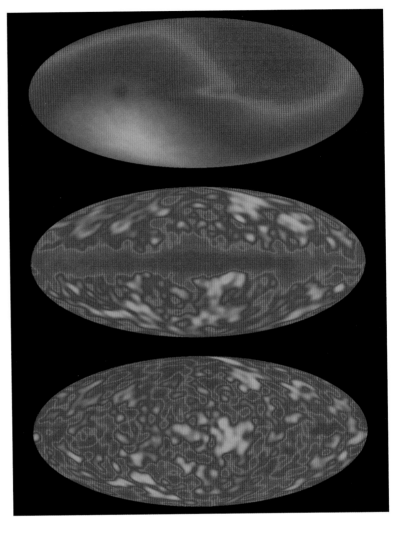

variations amounting to no more than one ten-thousandth of a degree. They may be
small, but these are the ancient seeds of the structures we see around us today. It may
sound strange to measure variations in density by measuring the temperature, but
there is a very good reason for it. As the COBE (COsmic Background Explorer) satellite
showed, matter at the time when the CMB radiation was emitted was not absolutely
uniform. Regions with a density above the average, gravitationally attracted still more
matter. The compression heated these regions slightly – and it is these variations that are
detected and measured.

Without any fluctuations for gravity to work on, the task of producing the non-uniform,
clumpy Universe we see today from a completely uniform Universe at the time of the CMB

would have been impossible. However, the dimensions of the fluctuations on the sky are also important. What our observations of the CMB produce is essentially a map of the whole sky, and it is easy to see that each of the blue (colder) and red (hotter) regions appear to be roughly the same size. They are on average about one angular degree across, about twice the apparent size of the full Moon. From this single piece of evidence, and some careful thinking, cosmologists can confirm that the Universe is flat. This is possible because we know the real, physical size of the fluctuations in the early Universe; they are predicted by theory. Comparing this expected size to the apparent size tells us how the light has been bent since it left its source, and that depends on the amount of matter in the Universe. The more matter there is, the more the light is bent. In a closed Universe the light would have been significantly bent, and the net effect would have been to make the fluctuations appear larger than expected. In an open Universe – one without much matter – the fluctuations would have appeared much smaller. In fact, comparing simulation to reality reveals that the Universe has just the critical amount of matter – and is flat.

This discussion illustrates a point that is the source of both excitement and frustration to cosmologists. Excitement, because it reveals that the study of the microwave background can tell us not only about the very early moment at which it was emitted but also about the entire history of the Universe since then. This is also a problem; if we want to draw firm conclusions about the early Universe, we must be careful to disentangle more recent effects, which can be difficult to do.

The barrier of light

We have seen that before the creation of the microwave background the Universe was opaque; no light could travel far through it. We can no more look back into this era than we can look up from the Earth and see the inside of a cloud. This analogy is not perfect because a cloud is not in itself luminous; a better picture is provided by the Sun. The Sun, viewed from outside, appears to have a definite surface (the photosphere), but what we are seeing is merely the boundary at which the material becomes transparent. Inside the photosphere the gas is so hot, luminous and dense, that no photons can pass through without colliding – similar to the state of affairs immediately following the Big Bang. Outside the photosphere the gas is transparent and photons can pass through freely, similar to what happened in the Universe immediately after the event of transparency – the moment when the CMB was created.

Looking through clouds on the Earth, we have a simple remedy – radio waves easily penetrate clouds and so we can still gain some information from beyond or within them. The same trick will not work with the CMB. The 300,000 year limit applies to all electromagnetic radiation, and seems to be an insurmountable barrier. How then can we talk with confidence, as we have been doing until the last few paragraphs, about conditions before then? For now, we have to rely on our theories, many of which are able to make predictions of how the microwave background will look. We can then compare these theories with the actual CMB, and draw the appropriate conclusions.

Ideally, however, we would like to be able to look back beyond this barrier, and there are numerous proposals as to how to achieve this. We may be able to detect highly energetic particles that have survived unchanged since before the microwave background epoch. We may be detecting such particles already, in the form of tiny, almost massless neutrinos or other exotic forms of matter.

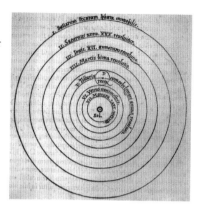

▲ **Copernicus's Universe**

This map by Copernicus was one of the first to show the Sun at the centre of the Solar System. A fundamental assumption underlying all of our attempts to understand the Universe is that there nothing special about our region of the Universe, and therefore we can draw conclusions about the whole from what we can see. We are guided by the so-called Copernican principle, which more formally states that no theory should place the observer in a special position. So far this has been vindicated – the Earth is not the centre of the Universe, and neither is the Sun. Neither of them is at the centre of our Galaxy, and our Galaxy is not the only one in the Universe, or even particularly distinguished.

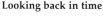

▶▶ Ice Fishing for Cosmic Neutrinos

Near the South Pole, astronomers have melted
around 100 holes and are using them as an
astronomical observatory: the IceCube Neutrino
Observatory. A string of football-sized light-detectors
sensitive to blue light is lowered into each hole,
and the water refreezes. Blue light is expected
from ice collisions with high-energy neutrinos.
Data from a preliminary experiment, AMANDA,
was used to create the first detailed map of
the high-energy neutrino sky, and IceCube will
be used to look for neutrinos coincident with
nearby supernova and distant gamma-ray bursts.

Looking back in time

Cosmologists may not be able to handle samples and submit them to analysis in the
laboratory, as chemists and physicists can do, but they have one tremendous advantage:
they can literally look back in time and observe the object of their study exactly as it was
millions of years ago. To see further and further back in time, remember, we need only
to look for objects that are further and further away from the Earth. As we have seen,
this does not apply to those events before the moment of transparency, which lie hidden
in the opaque infant Universe, but from now on we are discussing events that we could
potentially observe directly.

This chapter began at the moment the Universe became transparent, a moment we see
echoed as the cosmic microwave background. Follow-up experiments, from Earth and
in space, have confirmed COBE's detection of tiny variations in the temperature of this
radiation. These we interpret as an indication that there were indeed irregularities in the
density of the Universe at this point in time of about one part in ten-thousand. Yet the
variations in density we see around us today are much, much larger than this. We see
huge galaxy superclusters, regions where thousands of galaxies crowd together and other
areas of space that are almost devoid of matter.

Our own Milky Way Galaxy is just one of millions of spiral galaxies, and you might
imagine that there is no reason to doubt that the galaxies (or rather, the groups of
galaxies) are simply spread throughout the Universe at random. Yet large-scale surveys
of galaxies reveal a wealth of honeycomb-like structure on the largest scale, including a
'Great Wall' some 30 million light-years long. How did the Universe evolve from its early,
newly transparent, almost-but-not-quite uniform state to its present form?

Gravity, the universal force

The only force we would normally consider significant at astronomical distances is
gravity, and the strength of the gravitational pull of an object – whether it be a star, a
planet, a human being or a cloud of gas – depends on how much matter there is within
it. Note that 'mass' and 'weight' are not the same thing – mass is a measure of the amount
of material present, whereas weight describes the force due to gravity. Therefore an
astronaut in Earth orbit is weightless but is certainly not massless. We could define gravity
as being 'the force that gives mass weight'. For instance, the Moon is a relatively small
member of our Sun's family, and has such a weak gravitational pull that it has not even
been able to hold on to an atmosphere. The Earth is much more massive than the Moon,
and therefore has a greater ability to attract objects toward it. Thus, fortunately for us, it
can retain the atmosphere we breathe. Similarly, the dense regions in the early Universe
had a greater gravitational pull than the regions that were less dense, and so drew in

▼ How many scientists does it take to change
a light bulb?

Japan's Super-Kamiokande experiment uses many
thousands of photomultiplier tubes surrounding a
reservoir of ultra-pure water, to catch the flashes of
light from neutrino interactions in the water. In 2001,
most of the tubes exploded, necessitating large-
scale repairs. Here, scientists in an inflatable dinghy
check the tubes as the tank is refitted and refilled.

Neutrinos

These tiny particles have been studied by
astronomers for the last thirty years. Quantities
were produced a few minutes after the Big Bang;
others are a by-product of the reactions that power
stars, and are incredibly unreactive. In the course of
reading this sentence, the odds are that millions of
neutrinos from the Sun have passed through
your body without reacting, entered the Earth and
emerged from the other side of the planet. In order
to study them, astronomers and particle physicists
build detectors, consisting of vast tanks of liquid with
which the occasional neutrino might react. These
have to be built deep underground because on the
surface there would be too much contamination from

particles such as cosmic rays – high-energy nuclei that slam into our upper atmosphere at close to the speed of light, propelled across the Universe by the most powerful explosions known, of which more later. In their current form, these detectors are too small to be able to function as proper telescopes. They can tell us how many neutrinos are reacting with the detector, and measure their properties, but they cannot tell us from which direction in the sky the neutrinos are coming. For that, we will need much larger facilities. The largest and most sensitive is Ice Cube, which uses a cubic kilometre of the absolutely pure, transparent ice found under the South Pole as a vast neutrino detector.

▶▶ Virtual Universe

A still frame from a computer simulation of the development of the early Universe shows a slice one billion light-years across. Each filament contains the material that will clump to form thousands of galaxies, and the simulation shows that the Universe becomes more clumpy with time. This simulation includes the effect of dark matter, which interacts only via gravity. It does not, however, take into account the possible effects of 'normal' matter, which is a much more difficult computational task. Nevertheless, by comparing simulations like these with the observed reality, scientists have been able to learn a great deal about our Universe.

▼ Boomerang

The one-million-cubic-foot balloon that carried a CMB experiment into the stratosphere. It is shown here just prior to lift-off, with Antarctica's Mount Erebus in the background.

material from their surroundings. This, of course, further increased their gravitational pull – and so on, the process accelerating all the time. Here, as has often been the case, the rich got richer and the poor got poorer!

Inside each of these denser regions there were further localized variations in density, and the same sorts of processes operated – greater mass, greater pull, more runaway collapses. Using computers, we are now able to reconstruct what went on, and to build up a model which gives a good representation of the evolution of the large-scale structure that we see in the present-day Universe.

Wherever structures are being formed, we have to consider two opposing tendencies; the expansion of space, which began with the Big Bang and, locally, contraction under the influence of gravity. Once an object in the process of formation accumulated enough mass it was able to resist the overall expansion, and collapsed.

A typical ancestor of a galaxy cluster would initially have been small, growing in volume with the expansion of the Universe, all the time accreting matter from its surroundings. As it gradually ran out of matter to accumulate, it grew ever more slowly until its expansion ceased. The embryonic cluster of galaxies had reached its maximum extent and was then able to collapse to its final size. The force of gravity weakens with increased distance, and so at this stage in the evolution of the Universe, collapse was only possible on small scales – the first galaxies, still mere agglomerations of gas, were forming.

Gloomy times

What did these aggregations look like? We can't see them as we are still looking at what Martin Rees, the 15th Astronomer Royal, has called the 'Dark Ages'. During this period, which began immediately after the epoch of the microwave background, there were not yet any stars to light the Universe.

But there was, of course, the comparatively recent echo of the moment of transparency. This radiation (which perhaps at this point we should call the Cosmic Electromagnetic Radiation Background, instead of the CMB) started life at about 3000 degrees, around the temperature of an oxyacetylene torch, so there was in fact a diffuse glow, ever fading, and becoming redder, all through this period. In fact it may be true to say the Universe was never completely dark, just gloomy!

The gravitational collapse of the material that would eventually form galaxies continued in the fading light as the Universe cooled. Then came a dramatic change; the gloom was suddenly illuminated, when multitudes of stars burst forth. The Universe exploded in a blaze of light. How sudden this was is still a matter for debate, but in any case the time has come to consider the epoch of the first stars.

Balloon with a view

Before the development of space-based research, astronomers were severely handicapped. Ground-based instruments were simply unequal to the task of measuring variations in the temperature of the microwave background. The first results with sufficient resolution to view the variations were obtained only in 1992 by the satellite named COBE (COsmic Background Explorer). New data arrived in 1999, not from space but from balloon-borne equipment carried in a helium-filled balloon and taking advantage of the dry climate of Antarctica; some astronomers believe that the southern polar region may provide the best site for astronomical observation on Earth, and further testing is under way. There were two separate projects, Boomerang (The Balloon Observations of Millimetric

In the Big Bang itself, to all intents and purposes only three elements were created: hydrogen, helium and a smaller quantity of lithium; traces of other elements were negligible. All the other elements known to us today were synthesized inside stars. It has often been said that 'we are stardust', and this is true enough. The material in our Sun, and the entire Solar System has already been recycled through probably two previous generations of star formation. As we will see later, the explosive life history of many stars transforms the hydrogen and helium into heavier elements. The presence of

Extragalactic Radiation and Geophysics) and Maxima (Millimeter Anisotropy eXperiment IMaging Array). Boomerang had a main telescope with a primary mirror 1.2 m in diameter, and was carried by balloon up to a height of 37 km (23 miles); it covered an area of 1800 square degrees of the sky and produced a resolution some 35 times better than that of COBE. These images, which brought the microwave background into sharp focus, revealed hundreds of complex regions visible as tiny variations – each a difference of just 0.0001 degree.

Boomerang's images were surpassed by those from WMAP, the Wilkinson Microwave Anisotropy Probe, and later by ESA's Planck. These missions revealed tiny variations in temperature; the evidence of the earliest stage of galaxy formation.

▶ **Galactic census**

The Two-Degree Field (2dF) measured the redshift (equivalent to the speed they are moving away from us) of 240,000 galaxies. The experiment took measurements in two slices in opposite directions through the one billion light-year vicinity of the Solar System. These speeds translate into distances, and the resulting three-dimensional map reveals how mass is distributed in the Universe around us. It shows only between 20 and 30 per cent of the total mass needed for the Universe to be 'flat'. Since other experiments indicate that the Universe is flat, there is a clear need to speculate on the existence of some kind of dark matter – matter that makes its presence felt through gravity, but is not otherwise detectable.

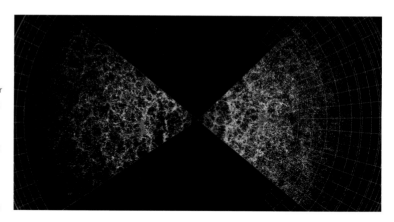

aluminium, for example, is a clear indication that some material came from a supernova explosion, while the Solar System's gold was partly produced in a collision between two exotic objects called 'neutron stars'. The first stars to form, on the other hand, were born containing only the three lightest elements.

In order to form a star, a parcel of gas must collapse, and to collapse it must cool. In the present-day Universe, radiation from carbon and oxygen atoms takes energy from the collapsing clumps of gas, but in the epoch we are describing – with no source of cooling but molecular hydrogen – the process is much less efficient. As a result, only large clumps can collapse and the stars that formed from them were also very large. The first stars were indeed extremely massive – perhaps as much as several hundred times the mass of our own Sun. With their huge reserves of fuel, one might expect these leviathans to shine for far longer than the Sun's lifetime, but in fact the opposite is true. The early stars lived fast and died young, actually surviving for just a few million years. By comparison, our Sun will have a total active lifetime of about nine billion years.

The source of stellar energy

To understand why this is, we need to consider the conditions deep in the centres of stars. Only one star is available for close study – our Sun. The Sun, like all normal stars, is a huge ball of incandescent gas, big enough to engulf well over a million globes with the volume of the Earth. Its surface is at a temperature of 5600 °C, while the core, where the energy is being produced, reaches around 15 million °C. We cannot look far into the Sun, but we can examine its constitution. We can develop mathematical models that seem to fit the observations, and so have confidence in our estimate of the core temperature. It contains a great deal of hydrogen, approximately 70 per cent of its mass. It is this hydrogen that is used as 'fuel'. And this is the same situation as in the first stars.

We have seen that a hydrogen atom, the simplest of all, has a single proton as its nucleus and one orbiting electron. Inside a star, the heat is so intense that the electron is stripped away from its nucleus, leaving the atom incomplete; the atom is said to be 'ionized'. At the star's core, where the pressure as well as the temperature is so extreme, these nuclei are moving at such enormous speed that when they collide, nuclear reactions are able to take place. Nuclei of hydrogen are combining to build up nuclei of the second

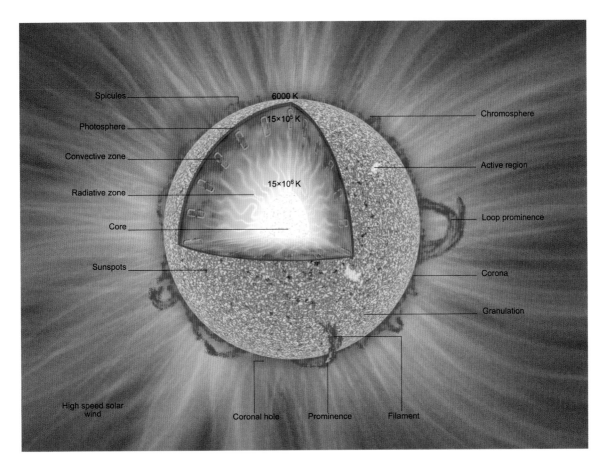

Spicules — 6000 K
Chromosphere
Photosphere — 15×10⁵ K
Convective zone —
Active region
Radiative zone — 15×10⁶ K
Core —
Loop prominence
Sunspots —
Corona
Granulation
High speed solar wind
Coronal hole   Prominence   Filament

▲ Inside the Sun

From the core to the photosphere is a distance of nearly 435,000 miles (700,000 km), approximately the same as a trip to the Moon and back.

lightest element, helium. Admittedly this takes place in a rather roundabout way, but in the end four hydrogen nuclei combine to make one nucleus of helium. There are also by-products; as well as the light we receive from the stars, there are strange particles called neutrinos, of which more anon. In the process of helium-building, a little mass is lost which causes a lot of energy to be released. It is this released energy that keeps the stars shining. The Sun is less massive now than it was when you started to read this paragraph. The supply of hydrogen fuel cannot last indefinitely, but there is no immediate cause for alarm. The Sun was born around five billion years ago, and as yet is no more than middle-aged by stellar standards. When all the available hydrogen has been used up, the Sun will not simply fade away; but that is another story to be told in another chapter.

So, in the Sun at least, it is the loss of mass in the conversion of four hydrogen nuclei into a lighter single helium nucleus that provides the energy that powers the star. The most famous equation in the world, $E = mc^2$, tells us that mass (m) is equivalent to energy (E). The converting factor (c^2), equal to the speed of light squared (multiplied by itself), is large, so that a tiny amount of mass loss produces a vast amount of energy. The Sun loses

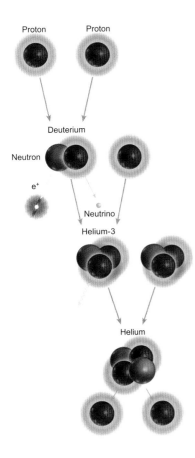

▲ Nuclear fusion

Hydrogen atoms fuse to form helium in the heart of the Sun, generating the energy that gives us light, heat and life. This process is known as the proton–proton cycle.

some four million tonnes of matter every second!

How does this disappearance of mass come about? Each of the four hydrogen nuclei is a single proton – hydrogen is the simplest of all atoms and consists merely of an electron orbiting a proton – whereas the helium nucleus is made up of two neutrons and two protons. However, a neutron is slightly heavier than a proton, so that if we just add up the masses of the particles on their own then it seems the helium nucleus must be heavier than four hydrogen atoms; mass seems to have been gained. Yet a helium nucleus really does weigh less than four protons, despite being composed of heavier particles. We are forced once again to remember that we are in the realm where quantum physics and its associated effects dominate, and the solution lies here. It is true that if we can measure the mass of a proton on its own it is slightly less than that of a neutron, but the subatomic particles are not free. In a helium nucleus they are bound together by the strong nuclear force and are less free to move. The creation of these bonds between subatomic particles releases energy and we measure a drop in the mass.

Why does the nucleus we produce have two protons and two neutrons? Life for astrophysicists attempting to study these reactions would be much simpler if it were possible to form a stable bond between two protons alone. This 'light helium' could then be produced by the direct, head-on collision of two protons, which would release electromagnetic radiation. However, the forces acting between two protons are not quite strong enough to hold them together when, as they both carry the same positive charge, the electromagnetic force is attempting to pull them apart. Instead of this simple picture of combining protons, the inside of the Sun, and indeed of all stars, is home to a subtle and a surprisingly slow process.

As we cannot simply add together two protons we must find a way to bypass this state, which blocks the formation of more complicated nuclei. In this discussion we only need to consider nuclei, not whole atoms, because at the temperatures that prevail at the centre of a star the electrons, which normally orbit the nuclei, making up an atom, have far too much energy to be captured. The only force that can help is the weak nuclear force, which spontaneously causes protons to decay into neutrons, releasing a positron and a neutrino. The newly-created neutron is captured by a passing proton, creating a deuterium nucleus. Deuterium is essentially heavy hydrogen, with a neutron added to the usual proton. The weak force lives up to its name, and this is the step that takes longest – a proton might spend an average of five billion years at the centre of the Sun before it is able to form a deuterium nucleus – but from here on in things speed up.

In an average of a second or so, the deuterium nucleus will snap up another proton forming a stable nucleus with two protons and a neutron – helium-3, a light form of helium. In an average of 500,000 years this nucleus will collide with another, forming the more familiar helium nucleus with two protons and two neutrons and releasing two protons, which start the cycle again. This step is delayed by the difficulty of forcing two large positively charged nuclei together. The strong force, which operates over extremely short distances, pulls the nuclei together, but they are repelled by the electromagnetic force, which keeps positively charged particles apart. Eventually the nuclei will pass close enough for the strong force to act, and we are left with energy in the form of radiation, a positron, which combines with its antiparticle and releases more energy, and a neutrino.

Neutrinos are tiny particles that move at high speed and rarely interact with other particles, and so they shoot out from the centre of the Sun relatively unimpeded by the

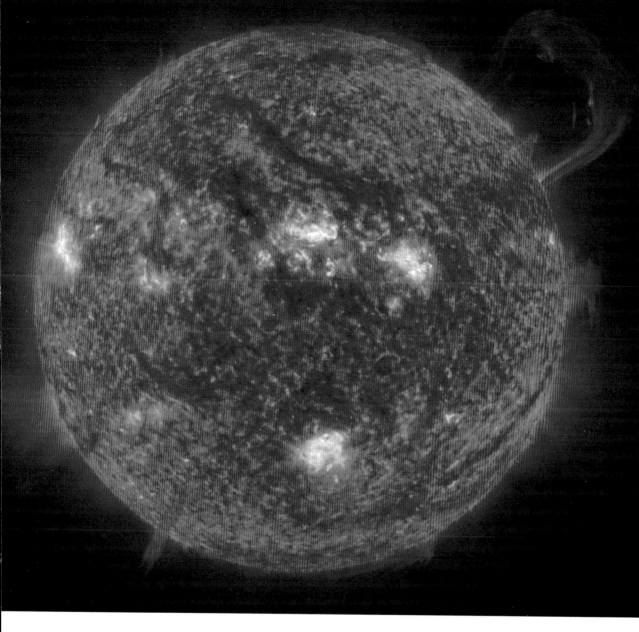

mass of gas surrounding the core. Some of them reach Earth, where vast detectors have been constructed to find them. For many years, this caused a problem as there were simply too few neutrinos being detected – it was thought one must be produced for each sequence of collisions that produced a helium nucleus. However, it turns out that the neutrinos have a remarkable ability to change 'flavour', or type, en route. Particle physicists know that there are three kinds of neutrino, and it turns out they have the ability to switch between kinds over time. The original experiments were sensitive only to one particular kind of neutrino, and therefore missed all the others. These experiments confirmed that our picture of what is going on in the centre of the Sun, at far higher temperatures than any experiment on

▲ The active Sun

This image shows a huge handle-shaped prominence – a cloud of relatively dense plasma suspended in the Sun's corona at a temperature of 60,000 degrees. The hotter areas are white, the cooler dark. In some cases these loops break off as a coronal mass ejection, hurtling a billion tons of plasma out into space.

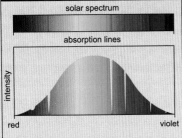

solar spectrum

absorption lines

intensity

red

violet

SPECTRA

Sir Isaac Newton was the first to pass a ray of sunlight through a glass prism, and to realize that the light was a medley of wavelengths from red (long) through to violet (short). He passed the sunlight through a hole and a prism, and drew it out into a coloured sequence – the first spectrum intentionally produced. Newton never took this experiment much further (possibly because the lenses available were of poor quality glass, but also, no doubt, because he had other things on his mind), and the next real development was due to the English scientist W.H. Wollaston, in 1801. Wollaston used a slit rather than a hole in his screen, and the spectrum of the Sun showed as a coloured band crossed by dark lines. Wollaston believed the lines simply marked the boundaries between the different colours – and thereby missed the chance of making a great discovery. The man who did so, more than ten years later, was the German optician, Joseph von Fraunhofer.

Like Wollaston, Fraunhofer produced a solar spectrum. He mapped the dark lines and found that they did not vary either in position or in intensity; for example there were two very prominent dark lines in the yellow part of the band. What caused the lines? The answer was given in 1858 by Gustav Kirchhoff and Robert Bunsen, who may be said to have laid the foundations of modern spectroscopy.

Just as a telescope collects light, so a spectroscope splits up light into its full spectrum, very much like a rainbow. Examine the spectrum of a luminous solid or liquid, and you will see a continuous band of rainbow colours. But a spectrum of a gas under low pressure will be quite different; instead of a rainbow there will be isolated bright lines – an emission spectrum (see right). Kirchhoff and Bunsen saw that each line was the trademark of one particular element or group of elements and cannot be duplicated. Thus sodium yields two bright yellow lines as well as a host of others. Some elements have complicated spectra. Iron, for instance, has thousands of lines. But the great insight was to realize that the dark lines they saw crossing the continuous spectrum of the Sun corresponded exactly to the bright emission lines emitted by glowing gases in the laboratory. We now know that each spectral line is generated by a particular transition in the state of an electron in the shell of a gas atom. If the gas is hot, we see an emission line, as the electron drops down an energy level, emitting energy, and if the gas is cool, viewed against a bright continuous background like the Sun, we see a dark absorption line, since the electrons are moving up a step in energy level, and absorbing energy at this same frequency. That distinctive pair of dark lines in the yellow part of the Sun's spectrum is a clear signature of the presence of relatively cool sodium gas. From a study of these Fraunhofer lines it has been possible to establish the abundance of all gaseous elements in

▲ Absorption spectrum

This picture illustrates the appearance of absorption lines. The intensely hot surface (photosphere) of the Sun emits white light, which passes through the slightly cooler outer regions (the chromosphere, featuring some solar prominences is seen here). This light, being split up into its constituent colours, or frequencies, by the prism, is revealed to be made up of a continuous humped-back 'black body' spectrum, typical of an incandescent object, crossed by the dark 'Fraunhofer' lines, evidence that the gases in the Sun's cooler layers have removed these particular frequencies from the picture.

▶ Newton's sketch

A reproduction of an original sketch by Sir Isaac Newton, showing the layout of his famous experiment to split white light into its component colours.

spectrum no.

1
2
3
4
5
6
7
8
9
10
11

the Sun's inner atmosphere, a region often referred to as the 'reversing layer'.

The dark lines, now called Fraunhofer lines, can give information about motion and, indirectly, distance. Listen to an ambulance sounding its siren. When the car is approaching, more sound waves per second reach the ear than would be the case if the car was stationary; the wavelength is effectively shortened, and the note of the horn is high-pitched. When the car has passed by, and has started to recede, fewer sound waves per second reach you, the wavelength is lengthened and the note drops. This is the Doppler effect, named after the Austrian who first explained it. Exactly the same thing happens with light. For an approaching source, the shortened wavelength makes the light more blue; with a receding source the light is reddened. The colour change is too slight to be noticed, but the effect shows up in the Fraunhofer lines. If all the lines are shifted towards the red, or longer wavelengths, the source is receding. The greater the redshift, the greater the velocity of recession.

Now let us return to the solar spectrum. The Sun's bright surface, or photosphere, gives a continuous spectrum. Above it is a layer of gas at much lower pressure (the chromosphere) and this might be expected to yield an emission spectrum. In fact it does so, but seen against the rainbow background the lines are 'reversed', and look dark rather than bright. The positions and intensities are not affected; the two dark lines in the yellow part of sunlight correspond to the emission lines of sodium, and so we can prove that there is sodium in the Sun.

▲ Historic spectra

This frontispiece from Norman Lockyer's 1874 textbook *Elementary Lessons in Astronomy* (never bettered!) neatly illustrates the correspondence between emission and absorption spectra. The two distinctive yellow sodium lines are here seen on their own in emission (spectrum 5) and in absorption against a continuous spectrum (6). They are also visible below as Fraunhofer lines in the spectra of Sirius (7), our Sun (8), and Betelgeux (9). The other lines in these spectra indicate the presence of many other elements.

▲ Central Milky Way

In visible light the centre of our Galaxy is obscured by dust. Viewed in the infrared, the dust actually glows, but the stars can be seen through it. Around one million stars are visible in this image, and the centre of the Milky Way lies in the glowing area at right.

Earth could hope to reach, is basically correct. They also provided the first firm evidence that neutrinos had a finite (although small) mass, for if they were, as had been believed, completely massless they could not switch from one kind of particle to another.

The life of the first stars

As the first stars to appear in the Universe – those whose light ended the Dark Ages – were massive, each perhaps matching the weight of as many as 150 suns, the increased gravitational pressures that came with their immense size heated their cores to very high temperatures. The nuclear reactions that power stars must have proceeded faster, and the material was used up rapidly. The first stars ran out of fuel in a period perhaps as short as a million years.

Before the birth of the first stars, the Universe was a sea of atoms, mainly hydrogen. The giant stars ignited, and their radiation spread outwards, knocking electrons out of atoms – ionizing them. Gradually, each new star was surrounded with a bubble of ionized gas. The more powerful stars would have produced larger bubbles. The star's energy could only influence the gas out to a certain distance, but these stars were large enough and energetic enough to create huge bubbles, tens of thousands of light-years across.

What happened next? Occasionally, the bubbles around two different stars met. As soon as they did so, all the matter within them was exposed to the combined light of the two

stars. Powered by twice the energy, the bubble expanded much faster and further. This meant that there was a greater likelihood that the expanded bubble would collide with another neighbour, and the whole process accelerated. Over a relatively short period, a Universe dominated by neutral hydrogen evolved into one in which more than 99 per cent of the material was ionized.

Black holes – a one way trip

There is another possible candidate for the cause of this first ionization. (Rather illogically, this period is known as 'reionization'.) Almost every galaxy, including ours, has a massive black hole at its centre. A typical black hole is the product of the collapse of a massive star. It has a gravitational pull so powerful that not even light can escape from it; its escape velocity is too large. The concept of escape velocity is straightforward enough; it is the velocity an object must attain to escape from the gravitational field of a more massive body. Eventually the escape velocity of a collapsing star rises to 186,000 miles (300,000 km) per second, the velocity of light. Light can no longer break free, and since light is the fastest thing in the Universe, the old star has surrounded itself with a forbidden zone from which nothing can escape. Obviously we cannot see it, because it emits no radiation at all, but we can locate it because of its gravitational effect upon objects that we can detect – for example when the black hole is one component of a binary-star system.

▶ **Escape velocity**

Throw an object upward, and it will rise to a certain
height, stop, and then fall back to the ground. Throw
it faster, and it will rise higher. Throw it up at a speed
of 7 miles (11 km) per second (admittedly, rather a
difficult thing to do) and it will never fall back; the
Earth's gravitational pull will not be strong enough,
and the object will escape into space, which is why
this value is known as the Earth's escape velocity.
The escape velocity of the Sun, a normal star, is
386 miles (618 km) per second, while the escape
velocity of the Moon, which has only an eightieth
of the mass of the Earth, is a mere 1.4 miles (2.4
km) per second. This is not high enough to hold
down an atmosphere; any air on the Moon has long
since escaped into space. (Actually, the Moon does
have an extremely thin atmosphere; it is continually
replenished with dust from the surface and
continually lost.) To escape from the Earth the Apollo
astronauts required a massive Saturn V rocket,
whereas to escape the Moon they only required the
small engines on the lunar module, as seen here.

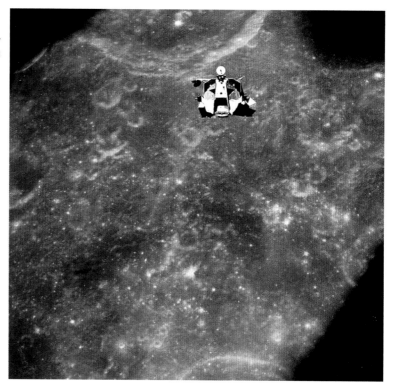

The result is that the black hole is cut off from its surroundings, and since no radiation
can escape we have no way of probing the interior. We can only speculate about
conditions inside. Falling into such a body would certainly be a one way trip and is
emphatically not to be recommended; scientists have coined the word 'spaghettification' to
describe this process – warning enough for anyone tempted to try a visit.

A black hole is usually produced by the collapse of a star, eight or more times the
mass of our Sun, but this may not be true for the very large black holes in the centres
of galaxies, which contain the equivalent of millions of solar masses. It may well be that
these very massive black holes formed at an extremely early stage of the Universe. If this
is so, then the first light may have come not from stars, but from matter heating up as it
as it fell into these black holes, and this would have been sufficient to cause widespread
ionization. In this case, the black holes responsible are then still with us, embedded in
the centres of today's galaxies. It is not yet clear which of the two possible mechanisms of
reionization is actually responsible. We need to learn a great deal more about this curious
epoch before this argument can be settled.

Supernovae

Whichever theory is correct, at some point these first, curiously large stars existed, and
their influence on their surroundings did not end at the time of reionization. We have

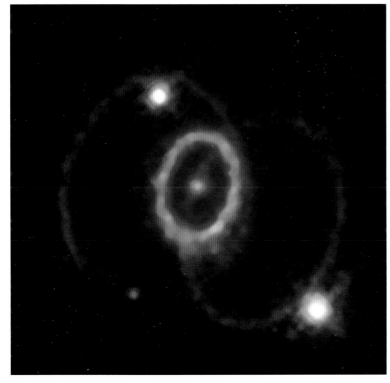

◀ **Supernova rings**

Astronomers are still waiting for the first observable supernova in our Galaxy since the invention of the telescope. In supernova 1987a we had the next best thing – a supernova in the neighbouring Large Magellanic Cloud. Seven years after the event, the Hubble Space Telescope imaged three extraordinary rings around the site of the explosion.

already seen that they led very brief lives; moreover, their deaths were violent. Unlike the relatively quiet future that awaits our Sun, such a massive star may be destined to suffer a cataclysmic explosion.

The outer layers of a star are supported by the energy produced in the nuclear reactions taking place in its core. When the fuel for this process is exhausted, these outer layers collapse, increasing the pressure and the temperature of the core. These changes allow helium nuclei, the product of the previous set of reactions, to collide and react with each other to build up heavier elements. Meanwhile, hydrogen around the core will still be being burnt; the result is rather like an onion with many layers, as successively heavier elements are produced in the core. Eventually iron is produced, and here the cycle stops. The nuclei of iron are the most stable of all, and therefore when they collide energy is lost rather than produced. Once a massive star forms a core of iron, nothing can prevent the outer layers from crashing inwards. A dense core quickly forms, and a shock wave rushes through the star, propelling the rest of the material outward in a vast explosion of heat and light – which we see as a supernova.

Supernova outbursts are certainly very violent. Even more extreme are hypernovae, which are created in very much the same way but involve exceptionally massive stars. Yet we have not yet witnessed the ultimate: the most catastrophic phenomena we know are called gamma-ray bursts.

▲ Supernova 1987a

Left, before the explosion on February 23, 1987, and right, 10 days after. A single supernova can outshine an entire galaxy.

Gamma-ray bursts

Gamma rays are the most energetic form of electromagnetic radiation, and have very short wavelengths – shorter even than X-rays, at wavelengths below 0.01 of a nanometre (a nanometre is one billionth of a metre). Although there is a more or less uniform background glow in gamma rays that is constant all over the sky, a few discrete sources are found. These sudden bursts of gamma rays, lasting up to a few minutes, are extremely powerful and can be seen right across the visible Universe. The initial burst of gamma rays is followed by an 'afterglow' in other regions of the spectrum, and the identification of this fading 'smoking gun' was the key that allowed astronomers to determine the distance to the more recent bursts; we now know that the gamma-ray bursts are indeed very remote.

The gamma ray story

At the height of the Cold War, military satellites were launched to look for sudden bursts of gamma rays, which are one of the signatures of nuclear testing. The American satellites sent up for this purpose did detect bursts, although they were not in the least what had been expected. The bursts lasted for anything up to a few minutes and sometimes no more than a few seconds.

All that could be found out about them was that they seemed to be distributed evenly around the

The power emitted in a single burst is almost unimaginable – during its entire lifetime the Sun will not emit as much energy as a single burst will manage in a few minutes.

It seems that, although different bursts may have different causes, at least some gamma-ray bursts are produced by the deaths of the most massive stars. Remember that once such a star has run out of fuel to power nuclear reactions, the radiation emitted from its core is switched off, and gravity finally wins the battle. The outer layers of the star rush inward, and the central regions collapse completely to form black holes. The outer layers, meanwhile, rebound and are thrown outward at tremendous speed. The energy is so great that the atomic nuclei, assembled during the star's lifetime, are ripped apart and briefly everything reverts to hydrogen. However, the energy available in this massive explosion

▲ **Crab Nebula (M1)**

This is the famous remnant of a supernova that exploded in AD 1054, observed by Chinese astronomers. Within this nebula lurks a spinning neutron star, which is all that remains of the star's core.

sky rather than being situated at one geographical location, which thankfully ruled out a nuclear test as the origin. For many years, it proved impossible to decide whether the bursts were weak, and therefore nearby, or extremely powerful and therefore extremely distant. It is now believed that these bursts emanate from sources around a billion light-years from us, and are incredibly powerful – probably the biggest 'bangs' since the Big Bang itself.

▶ **Gamma-ray bursts**

A whole-sky map of the 2704 gamma-ray bursts that were recorded over a period of nine years by the Compton Gamma-Ray Observatory. The plane of our Galaxy runs horizontally along the centre of this representation, from +180 to –180.

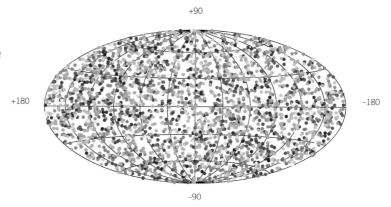

can then drive further nuclear reactions, which fuse the hydrogen atoms into heavier elements, including, significantly, those more massive than iron.

When the star involved is as large as those believed to make up the first generation, this outpouring of energy will be great enough to power a gamma-ray burst. In the nearby Universe, where the largest stars are only 20 to 30 times the size of the Sun, we see their deaths as relatively modest supernovae. The light from a single supernova, however, is still enough to outshine the entire galaxy in which it lies, and so hypernovae should be visible from right across the observable Universe.

Following this violent death, a shock wave rippled out from the explosion at close to the speed of light. A similar process can be seen in Hubble Space Telescope images of nearby supernovae. As well as heating the surrounding gas, the spreading shock wave from a dying first-generation star caused surrounding gas clouds to collapse in turn, triggering the formation of the next generation of stars. As these new stars were forming, they accumulated elements produced in the first generation of stars, which had not been available in the earlier period. These atoms, particularly carbon and oxygen, efficiently radiated away energy from the collapsing cloud. This allowed it to cool and fragment, producing smaller clumps, and eventually, smaller stars. Consequently, these second generation stars were very like those that we see today. The smallest of them (and hence those with the longest lifetimes) may even still be shining today, and we may well have detected them in our own Galaxy.

The exact mass of these stars had a profound effect on their fate. For example, stars with masses greater than about 300 solar masses would collapse directly to form massive black holes with no material expelled and no spreading shock waves. A star with a mass in a narrow band around 160 solar masses produces what is known as a pair-instability supernova. These explosions happen to produce vast numbers of positrons, the antiparticle of the electron. When particle and antiparticle meet, they annihilate, producing energy, and in these supernovae this energy is great enough to prevent the core collapsing. No black hole or neutron star is formed, and all the material is thrown outward. We believe large numbers of stars of this size may have formed in the early Universe, which is just what is needed to produce the material which will fuel the rapid formation of a second generation of stars.

▶▶ **Supernova blast wave**

This Hubble photograph shows a small section of the Cygnus supernova blast wave some 2,400 light-years away. The wave is expanding at 220 miles (350 km) per second.

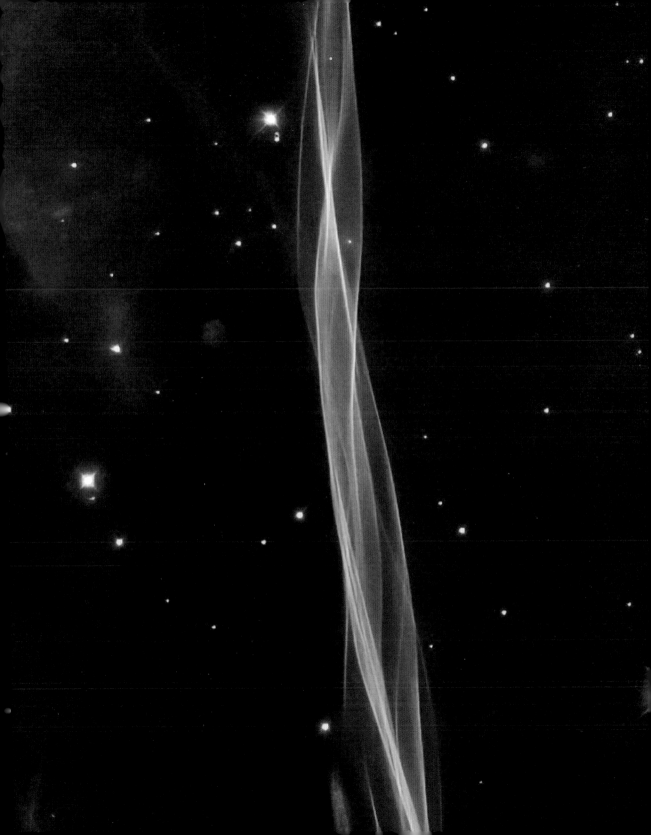

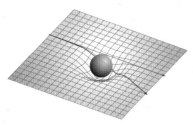

▲ Imagining spacetime

A massive object distorts space and time.
A ray of light is bent, so what we see
appears to come from somewhere other
than where it actually originated. The
dotted line shows the path undisturbed
light would take. With a massive object in
the way, the light follows the path shown
by the unbroken red line.

Relativity – an observer's guide

The physics of black holes is naturally written in the language of the General Theory of
Relativity, and it is worth taking the time to try to learn some of this language. According
to Einstein, if two different observers, each with their separate frame of reference, are
accelerating (or decelerating) relative to each other, their timescales will not agree. In
other words, while I may observe ten seconds elapsing, you, who are accelerating away
from me, may observe only six.

The temptation is first of all to ask who is right, and then to look for some subterfuge
that may have altered the clocks. Yet relativity tells us firmly that both are right and there
is no trick – different observers really do experience time flowing at different rates. Some
rules of common sense are preserved; two observers will always agree on the order of
events, for example. So although one may believe A preceded B by a minute, and another
that A and B were simultaneous, it is impossible for any observer to see B preceding A.
Hence cause and effect are preserved, but many other common-sense ideas that seem
second nature to us must be abandoned.

Why are such seeming paradoxes not part of our everyday experience? We never
notice clocks running at different rates, after all. The answer is that, fortunately, we don't
live anywhere near a black hole. Without extreme accelerations, or huge velocities near
to the speed of light, or very large concentrations of mass, the effects are so small that
Newton's laws of motion still work very well. Einstein did not prove Newton was wrong,
he extended Newton's ideas to be accurate in these more extreme cases.

So much for the effect of the black hole on the passage of time, but relativity also
tells us how its immense mass affects the space around it. One of the reasons relativity
is difficult to understand is that its mathematics is framed in a four-dimensional form
– the three familiar dimensions of space plus one of time. Space and time no longer
exist independently – Minkowski, who provided much of the mathematical structure of
relativity, went so far as to write that 'space by itself, and time by itself, have vanished into
the merest shadows, and a kind of blend of the two exists in its own right.'

Can you imagine what a four-dimensional sphere looks like? Neither can we, but
we can get some idea of its properties by considering just two dimensions, picturing
spacetime as a flat sheet of bed linen, being held taut at its four corners. Now, put a
tennis ball or some other weight in the centre, and the sheet will be distorted, just as
the theory tells us massive objects distort space and time. A light ray travelling in this
distorted space time will have its path distorted and bent. Around a massive black hole
this effect may be large enough to allow a suitably placed observer to see the front and
back of the surrounding disk simultaneously.

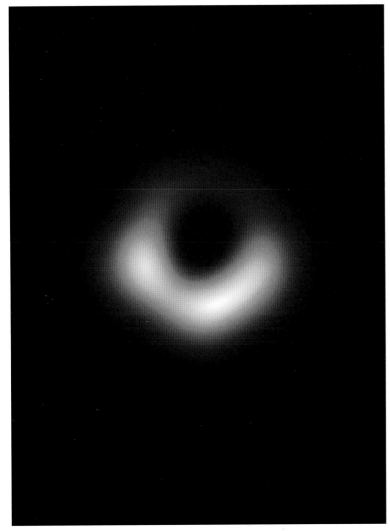

◀ **Supermassive black hole**

The supermassive black hole at the core of supergiant elliptical galaxy M87 has a mass about seven billion times that of the Sun, seen in this first false-colour radio image from the Event Horizon Telescope, which combined data from telescopes around the world to produce this single image.

700 MILLION TO 9 BILLION YEARS A.B. (AFTER THE BANG)

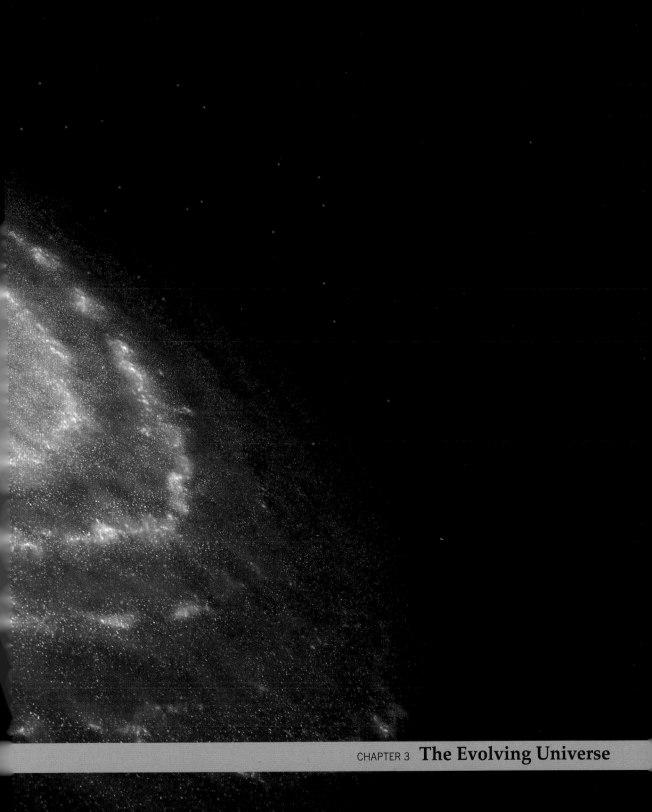

CHAPTER 3 **The Evolving Universe**

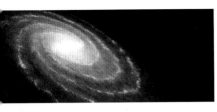

▲ The Milky Way

This artist's impression of our own Milky Way Galaxy shows the blue spiral arms, full of newly formed and newly forming stars, stretching out from the yellow bulge that surrounds the nucleus. Recent observations suggest that this bulge conceals a central bar, but the position of our Solar System within one of the spiral arms makes it hard to be sure of this.

▶▶ The Hubble Extreme Deep Field

Galaxies seen at a distance of 13 billion light-years when the Universe was just a few per cent of its present age. Combining ten years of data, the Extreme Deep Field is the deepest view into the Universe yet, revealing some 5,500 galaxies in a tiny patch of the southern skies, just a fraction of the angular diameter of the Moon.

▶ Quasar 3C175

The term *quasar* (short for quasi-stellar object) was coined to describe a group of surprisingly energetic point-like sources of radiation that were detected in the 1960s. Such an object is now more properly identified as an extreme example of an active galactic nucleus, known to be powered by a massive black hole. Even the centre of our own Milky Way Galaxy is active to a degree, but quasars are millions of times more energetic and are only seen far away, in the Universe of the distant past. This radio telescope image shows a quasar (the central dot) emitting streams of particles (only one jet is clearly visible), which reach to over one million light-years from the centre. When the jets of particles, travelling at near the speed of light, hit the surrounding gas, they create the two blob-like shock fronts.

After two chapters, we have finally arrived at a point in the history of our Universe where there are discrete objects that we can actually see. Even before the advent of the first stars, the collapse of matter to form galaxies was in progress, and the Hubble Space Telescope's deep field images reveal galaxies just 700 million years after the Big Bang. They are not like the systems we see around us: many are smaller, and there is a wide variety of weird and wonderful shapes on view. Some already harbour massive black holes. These are the mysterious quasars that dominate the scene. These powerhouses are now known to be the cores of very active galaxies shining with a luminosity equivalent to that of several thousand Milky Ways. Because they are so luminous they can be seen across vast distances, and all date back to the Universe at a fairly youthful stage.

Supermassive black holes

In the centres of these galaxies, even early on, lurked supermassive black holes amounting to several million solar masses. These could be the ones that formed directly from collapsing gas as we discussed earlier, or they may be the remnants of massive stars that have since swallowed a huge amount of extra material. In either case, a black hole this size has a gigantic gravitational pull and can attract vast amounts of material.

It seems that in the early stages of galaxy formation there was a huge amount of gas and dust available as star formation was just getting started. This material fuelled the black hole, and spiralled inward forming a disk. As it did so, it emitted light channelled into jets so that when we look down the throat of one of these jets, we see the powerful beacon we call a quasar. In this early period of the evolution of the Universe, collisions between embryo galaxies must have been common; as the two systems merge, fresh material is driven toward the black hole (or, indeed, black holes) and the quasar shines on. In fact, it may well be that all massive galaxies, including our own, went through a quasar stage in their evolution, and several quasars studied recently appear to be otherwise normal galaxies. Eventually, the fuel ran out, and the galaxies settled down.

We can glimpse this era by looking back to some of the earliest galaxies ever detected, which are shown in an image known as the Hubble Extreme Deep Field, that was obtained by

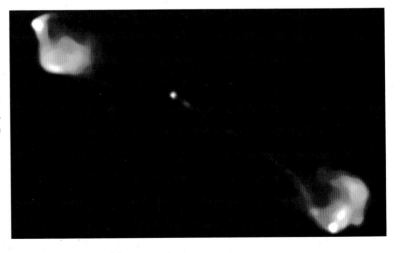

▲ Centaurus A

An elliptical galaxy 10 million light-years from Earth. This mosaic of Hubble Space Telescope images taken in blue, green and red light has been processed to present a natural colour picture. Infrared images have shown that hidden at the centre are what seem to be disks of matter spiralling into a black hole with a billion times the mass of the Sun. Centaurus A is apparently the result of a collision between two galaxies, and the left-over debris is being consumed by the black hole.

using the Hubble Space Telescope. Initially, for just over eleven days, the orbiting observatory was pointed at a patch of sky that had previously appeared to be completely devoid of any interest. This extremely long exposure allowed the light received from even the faintest objects to build up to detectable levels, and enabled the telescope to transform the blank patch of sky into one teeming with literally thousands of galaxies. This was called the Hubble Deep Field. Over the next ten years further photos of this patch of sky were assembled to create the most detailed ever image of the distant Universe – the Hubble Extreme Deep Field. Each speck in this image represents not a background star, but a background galaxy, and although a few are relatively nearby, and look completely normal, most of them are much smaller, fainter, and frankly odder. Even by eye one can draw initial conclusions from the image; for example, galaxies that appear red are the most distant, due to their huge redshifts. Hence we can begin to put the objects detected into a rough evolutionary sequence.

From looking at these earliest galaxies and attempting this kind of analysis, we can gain an insight into the formation of the galaxies we see today. We no longer believe that each galaxy formed in isolation; if this were the case the Extreme Deep Field would have shown a smaller number of larger, 'normal' galaxies. The new picture, initially suggested by simulations, is of early collapse leading to small structures, which merge through a series of collisions to produce larger systems. The detection of the fuel for this process in the form of the vast numbers of small galaxies in the most distant regions of the observable universe adds weight to this theory. What we may be seeing are the building blocks that make up the more familiar modern-day galaxies. This process continues today; in recent years we have

realized that the Milky Way is a cannibal, ripping several dwarf galaxies apart.

These small systems orbit the larger galaxy, but gradually get pulled inwards. Eventually, their orbits have been distorted to the extent that they regularly pass through the disk of the larger galaxy, and on each pass their gas and dust is stripped by the larger system. After several such encounters, the smaller galaxy becomes part of the larger system – the fate that awaits the two most obvious companions of the Milky Way, the Large and Small Magellanic Clouds.

These two satellite systems of the Milky Way have already been disrupted by their passage through the plane of our Galaxy's disk, as shown by a stream of stars which trail behind them. Recent observations made by the European Space Agency's galaxy-mapping Gaia satellite suggest the Large Magellanic Cloud (LMC) at least is having a reciprocal effect on the Milky Way. Our Galaxy appears to be lopsided, its disk sloshing under the effect of the LMC's gravitational pull. The structure of the Galaxy we call home is still altering under the influence of mergers rather like those which happened more frequently in the early Universe.

The exotically coloured galaxies in the beautiful Extreme Deep Field image (see page 73), which is likely to remain unique until the advent of Hubble's successor, remind us in a spectacular way of the principal evidence we possess of the central premise of this book – that our Universe is indeed expanding. The different colours of these myriad objects indicate different red shifts; the redder the object, the faster it appears to be receding from us. The light we see left them only 700 million years after the Big Bang – just five per cent of the age of the Universe. This has been verified from analysis of the position of spectral lines in these galaxies, examined in studies using ground-based telescopes.

▲ **Hidden companion**

On the far side of the central Milky Way, a small galaxy, known as the Sagittarius Dwarf Spheroidal Galaxy, has been found. It was once our nearest galactic neighbour, but is now being ripped apart by the powerful gravity of the much larger Milky Way. Peering through the Milky Way's stars, it was noticed that some of the background stars were not moving as they should. The galaxy is roughly the shape of the red region, and is a mere 80,000 light-years away. Our illustration shows a contour map of the radio wave intensities recorded for this galaxy, superimposed on a visible-light photograph of the region.

▲ **Magellanic Clouds**

The Large and Small Magellanic Clouds, visible only to observers in the southern hemisphere, are our second and third closest neighbouring galaxies at 179,000 and 210,000 light-years away, respectively. They orbit the galactic centre, and it seems they pass through the Milky Way's disk regularly, losing stars each time.

Throughout this period, structures were still forming from matter collapsing under its own gravity, just as they had been in the Dark (or Gloomy) Ages. Among them must have been the seed that would lead to the Milky Way Galaxy, a system that is rather above the average in size, though not exceptional; its quota of 100 billion stars is exceeded by the neighbouring Andromeda Spiral. Neither is the local group of galaxies exceptional; other groups are much more populous. The Virgo Cluster, whose members are on average around 60 million light-years away, contains well over a thousand large galaxies.

Discovery of spiral galaxies

One of the most important discoveries in the history of astronomy was that the misty-looking objects once called 'spiral nebulae' are other star systems well outside our Galaxy and that they are racing away from us in the general expansion of the Universe that began at the start of time and is still going on. The rate of expansion has not always been the same.

The discovery that many of the galaxies are spiral in form was made by the third Earl of Rosse, an Irish nobleman. At Birr Castle in County Offaly he constructed a reflector with a metal 72-inch mirror – much the largest ever made up to that time – and used it to examine the nebular objects; his drawing of the Whirlpool Galaxy, M51 (left) is amazingly accurate, though it was made as long ago as 1845. For many years the 72-inch reflector was out of use, but it is now in full operation again. Nothing like it had ever been built before, and nothing like it will ever be built again!

Our Galaxy, the Milky Way

Young galaxies contained large reservoirs of gas and dust that could be converted into stars. They are likely to have been dominated by the light from bright, young, blue stars and to look somewhat like our own Galaxy – a perfectly normal spiral. It is worth taking a slightly more detailed look at the Milky Way Galaxy before turning to others. We know it to be spiral in form, and we know that the galactic centre is about 27,000 light-years away from us. The overall diameter of the system is over 100,000 light-years, and in shape it has been likened to a double-convex lens (in less scientific terms, two fried eggs clapped together back to back). Looking along the plane of the system, we see many stars in almost the same line of sight, causing the lovely band of light which traverses the night sky, known since time immemorial as the Milky Way. The diameter of the central bulge (the yolks of the fried eggs) is around 20,000 light-years. Out of the plane and away from the main disk are the vast, condensed globular clusters as well as many 'stray' stars, which inhabit what we call the 'galactic halo'.

We cannot see the galactic centre easily, because there is too much obscuring material in the way, but this is no obstacle to radio waves and X-rays. The centre lies behind the star-clouds in Sagittarius. The exact location is marked by an intense radio source known as Sagittarius A* (pronounced Sagittarius A-star). In the central region there are swirling dust-clouds, plus clusters of very powerful stars, and very near the true centre there is

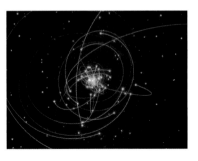

▲ **Galactic centre**
This simulation based on observations made by the European Southern Observatory (ESO) shows the orbits of stars very close to the supermassive black hole at the heart of the Milky Way. By following the motions of stars orbiting the black hole astronomers have been able to determine its mass. This is also a perfect laboratory to test gravitational physics and specifically Einstein's general theory of relativity.

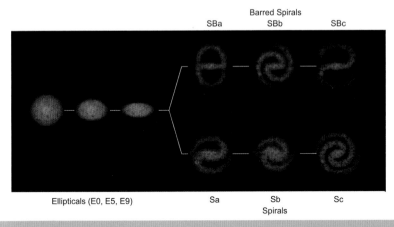

Barred Spirals
SBa　SBb　SBc

Ellipticals (E0, E5, E9)　Sa　Sb　Sc
Spirals

▲ **Classification of galaxies**

It was Edwin Hubble who drew up a system of classification for galaxies. The result is popularly called the Tuning Fork diagram, for obvious reasons – see above). Some galaxies are elliptical, while others are spirals, like Catherine wheels, and yet more are irregular.

The elliptical galaxies range from E0 (virtually spherical) through to E9 (very flattened). Spiral galaxies may be Sa (tightly wound), Sb (looser) or Sc (looser still). Some spiral galaxies show a bar

through their major axes, with the spiral arms extending from the ends of the bar (SBa, SBb or SBc). Many readers may be pleased to know that it is now thought that our own Milky Way Galaxy has a bar!

It was once thought that the Tuning Fork represented some sort of evolutionary sequence, with an elliptical galaxy turning into a spiral one or vice versa. However, the picture is now known to be much less straightforward than this.

▲ Sombrero Galaxy (M104)

Named for the famed Mexican hat, this galaxy's hallmark is the dark dust lanes comprising its spiral structure. We view it almost edge-on, only six degrees north of its equatorial plane.

▶▶ Clockwise from the top: Whirlpool Galaxy (M51)

This exotic object comprises a large spiral galaxy and a smaller, barred and more amorphous companion. The perfect spiral arms of the larger galaxy may be the result of the pull of the smaller one. It was the first of the external spirals to be resolved into individual stars.

Giant elliptical galaxy (NGC 1316)

Dust lanes are visible, and some of the star-like objects nearby are globular clusters – huge star systems of 10,000 to 1 million stars. Most of the stars in elliptical galaxies such as this are old, formed more than two billion years ago.

NGC 4622

Imaged by the HST, the galaxy NGC 4622 surprised astronomers by appearing to spin clockwise as opposed to the counter-clockwise direction predicted by simulations. Two of its outer arms are pointing towards the direction of rotation. Astronomers suspect that NGC 4622 interacted with another galaxy to cause this odd behaviour.

a black hole with a mass around 2.6 million times that of the Sun. The evidence for this comes from a very close look at 28 stars near the galactic centre. Since 1992, astronomers have watched them orbit whatever lurks at the centre, moving at orbital speeds of thousands of miles per second. By tracking the motion of these stars, we can calculate the mass and estimate the size of the central body. With so much mass packed into such a small space, the object at the centre of our galaxy cannot be anything except a black hole.

The Galaxy is rotating. The Sun takes about 225 million years to complete one orbit – a period often called the cosmic year. One cosmic year ago, the most advanced life-forms on Earth were amphibians; even the dinosaurs had yet to make their entry. (It is interesting to speculate about what our world will be like one cosmic year hence!) We travel not far from the main plane of the Galaxy, and we have just left one of the spiral arms, known as the Orion arm, so that we are now in a relatively 'clear' area.

Spiral galaxies

Many galaxies are spiral, their disks marked by beautiful curving spiral arms, sparkling with blue stars. These arms are now thought to be due to pressure-waves that sweep round the system. These are regions where the density of the interstellar material is greater than average, which triggers star formation. The stars that are most visible are very massive, and short-lived by cosmological standards before exploding as supernovae, but their brilliance makes the spiral arms dominant; when the pressure wave sweeps on, furious star formation stops, and the spiral arms become less evident. The sweeping pressure wave then creates a new arm. If this scenario is right, then in several tens of millions of years' time our Galaxy will still have spiral arms, but they will be made up of different stars.

The physics that governs the spiral arms of our galaxies can be likened to a more mundane problem – traffic jams. Consider traffic on the M25, the circular motorway around London. All the cars travel at roughly the same speed, but, if the road is busy, a car travelling at a slightly reduced speed can cause a build-up of traffic behind it. This

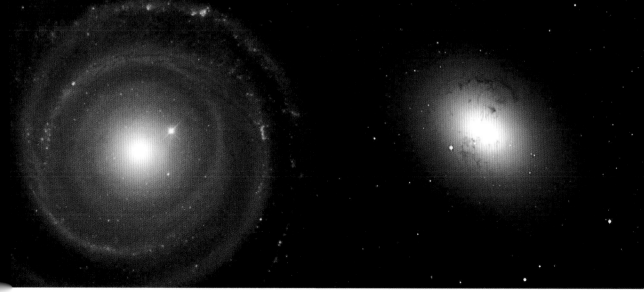

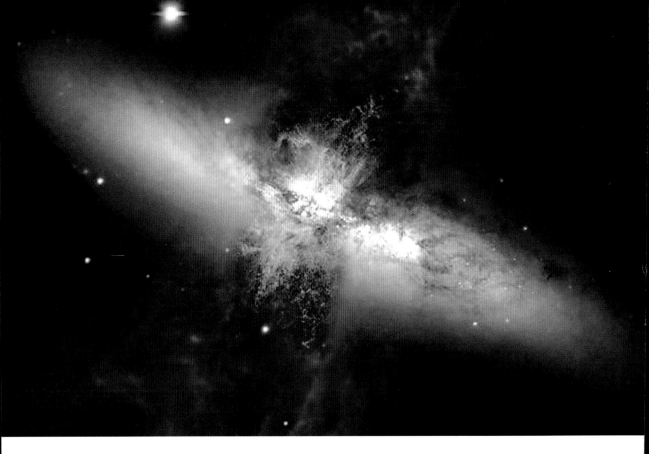

▲ Cigar Galaxy (M82)

What is lighting up the irregular Cigar Galaxy? It was disturbed by the close pass of nearby M81, and recent evidence indicates that the red expanding gas may be being driven out by the combined particle winds of many stars creating a galactic superwind. Material caught up in this wind is moving at enormous velocities.

▶▶ Penguin (NGC 2936)

This image shows the two galaxies interacting. NGC 2936, once a standard spiral galaxy, and NGC 2937, a smaller elliptical, bear a striking resemblance to a penguin guarding its egg. This image is a combination of visible and infrared light, created from data gathered by the NASA/ESA Hubble Space Telescope Wide Field Planetary Camera 3 (WFC3).

is exactly what happens to dust or gas orbiting the centre of a galaxy as it piles up in the spiral arms. Individual cars are part of the hold-up for just a limited time and will eventually move past it and on round the motorway, but the jam persists, comprised of other cars approaching from behind.

We have been able to measure the rotation of many spiral galaxies – mainly by the Doppler effect. If a spiral is rotating, then all the material on one side will be approaching us and all that on the other will be receding (when due allowance has been made for the overall motion of the galaxy, of course). This motion will be revealed by the position of the spectral lines and therefore the rate of rotation can be found. But there is one peculiarity that has profound significance.

Mysterious dark matter

In our Solar System, the orbital speed of a planet decreases with increasing distance from the Sun, because gravity becomes weaker further from the Sun. Logically, the same laws should apply to rotating galaxies. Stars near the centre should move much more quickly than those that are further out. Yet astronomers found, to their consternation, that this did not happen; far-out stars have a shorter cosmic year than they would be expected to, and so the spiral arms do not rapidly wind up. The galaxy seems to behave like a cross between the solar system and a solid body, acting like a spinning bicycle wheel. A speck

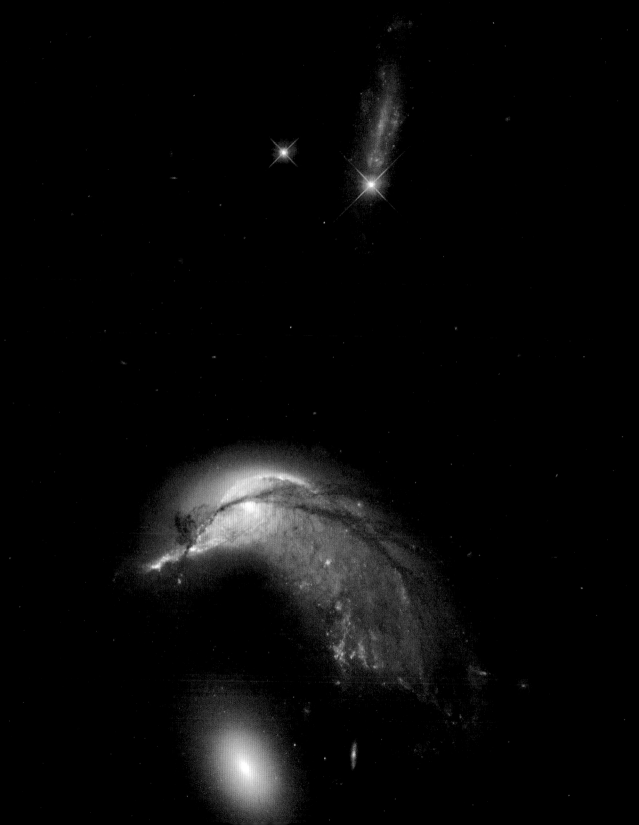

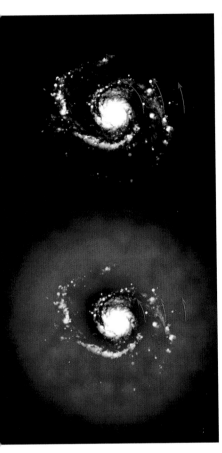

▲ Dark matter

If the matter we see is all that exists, the laws of
gravity predict that the outer arms of a galaxy will
rotate more slowly than those near the centre (top).
In fact, observations show that all the arms rotate
at the same speed (bottom). One explanation is the
presence of a halo of unseen dark matter.

of mud close to the hub will move more slowly than one on the rim, yet both complete a
revolution in the same time.

If the stars in a galaxy simply orbited a central mass like the planets orbit the Sun,
there would be no way of explaining this strange behaviour. The only possible answer is
that the mass of the system is not concentrated at or near the centre, but must be spread
throughout the disk and the outer parts of the galaxy. The most plausible explanation
is that there is 'dark matter' distributed throughout the galactic halo. The dark matter is
totally invisible but betrays its presence by its gravitational pull.

Can the dark matter be something ordinary, such as vast numbers of low-mass stars,
so faint that we cannot see them except when they are very close, at least by cosmic
standards? Certainly there are a great many stars (a very recent estimate gives the total
number of stars in the observable universe as 7×10^{24}) but it does not seem that their
combined mass would be nearly enough to account for the amount of dark matter.

Could the matter be locked up inside black holes? There would have to be an enormous
number of them, and we can look for them by monitoring a field full of distant stars for
the bending of light that would be produced by the passage of a black hole across the
field. Such 'microlensing events' in fact have been seen, but they are very rare indeed;
there cannot possibly be enough black holes to remove the need for dark matter. The
same is true for wandering planets amongst the stars, or any other form of Massive
Compact Halo Objects, or MaCHOs, which could serve the same function.

A solution that initially seemed much more promising involved neutrinos, which
are fast-moving particles with no electrical charge; they are not easy to detect but are
unbelievably plentiful, produced in vast quantities by the reactions that power the stars.
Many thousands pass through our bodies every second. If neutrinos had even a slight
amount of mass they could provide an explanation for the dark matter. We now know
much more about them than we did a few years ago, and though they are not completely
massless it seems certain that they cannot provide enough mass to solve our problem.

We are left with one solution. Dark matter may be composed of as yet unknown
fundamental particles, each of a small individual mass but existing in sufficiently large
quantities to explain the discrepancies. These hypothetical particles are known as
'WIMPs', or Weakly Interacting Massive Particles, and there are specific predictions from
particle physics as to exactly what they might be. We wait the discovery of a passing
WIMP – but there is worse to come when we encounter dark energy a little later.

Is there an alternative to dark matter?

Our aim here is to give an account of a single model of how the Universe is evolving,
supported by most of the currently available observational evidence. But in many ways the
model is deeply unsatisfactory, relying as it does on two as yet unknown components of
the Universe labelled dark matter and dark energy. Any theory which can dispose of the
need for such mysterious actors on the cosmological stage deserves to be taken seriously.

The most promising alternative theories fall under the general banner of 'MoND',
or Modified Newtonian Dynamics, and remove the need for dark matter by making
minor changes to the theory of gravity. MoND achieved significant success in the past,
accounting for the most of the evidence for dark matter on galactic scales. However,
recent observations of the galaxy cluster 1E 0657-56 – the Bullet Cluster – show effects
which are much harder to explain using simple MoNDian arguments.

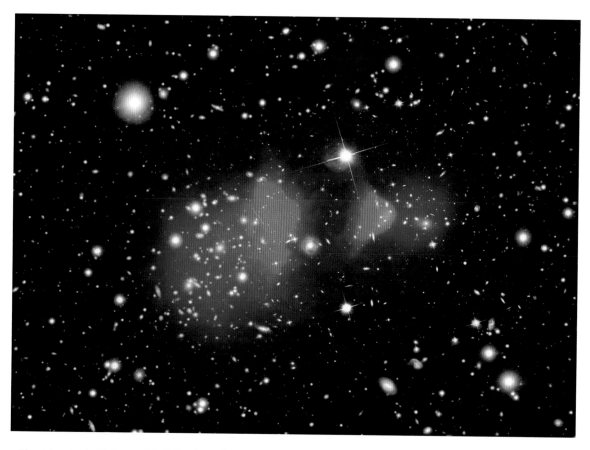

The pink region in this image of the Bullet Cluster (top right) is the X-ray glow of hot gas, detected by the NASA satellite Chandra. The blue regions represent an estimate of where the mass in the cluster lies, ingeniously traced by its effect on the light from background galaxies. The 'normal' matter, seen in X-ray radiation, is at the centre of the cluster, while the total mass is much more spread out. This can be explained only if the majority of the mass in the cluster is in the form of cold dark matter – the stuff that we have seen is necessary to account for galaxies having enough gravitational attraction to hold galaxies together. Here's how.

The Bullet Cluster is actually believed to be two galaxy clusters which have recently collided. Our illustration overleaf shows stills from NASA's animated movie, viewable at www.banguniverse.com, which convincingly portrays the way that this could have happened. Astonishingly, in a cluster of galaxies such as this, the gas in the vast spaces between the individual galaxies makes up roughly half the total mass. It is important to realise that in these pictures, at these wavelengths, we do not see the galaxies at all – we are 'seeing' only the gas. As these two clusters begin to pass through each other, the individual galaxies rarely, if ever, collide, because the spaces between them are so huge, but the gas molecules in the spaces do collide, scattering each other, and effectively slowing down the movement of this material.

▲ **The Bullet Cluster**

In this composite image of galaxy cluster 1E 0657-56, the individual galaxies are shown as they appear at optical wavelengths. Their total mass adds up to far less than the cluster's two clouds of hot, X-ray emitting gas shown in red. With more mass than the galaxies and gas combined, the distribution of the dark matter in the cluster is shown in blue. The dark matter was mapped by observing gravitational lensing of the background galaxies (see pages 87–9).

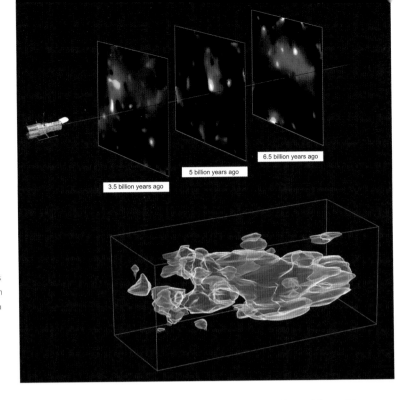

6.5 billion years ago

5 billion years ago

3.5 billion years ago

▶ **Seeing the invisible**

This 3D map is the first to show the distribution of dark matter in the Universe. It shows the filaments collapsing under gravity and growing clumpier with time. How can we see this invisible stuff? The map was constructed by looking at the shapes of half a million faraway galaxies: the dark matter deflects their light slightly, so we do not see them directly, rather through their influence on light heading towards us.

This is how 'normal' matter behaves, and the result is a large lump of normal (baryonic) matter, which we see as the central pink shape in the illustration, made up of the gas in both of the original clusters. However, if we suppose (as seems the simplest assumption) that the cold dark matter in the clusters interacts only gravitationally, there is in this case no slowing down due to collisions. Thus, as the whole process evolves, the dark matter in the clusters is able to keep moving through, and past, the agglomeration of gas. It ends up concentrated in the outskirts of the cluster, to each side of the centre, exactly what we see in the shape of the 'Bullet'.

The method used to work out the distribution of mass in the Bullet Cluster has been used for other clusters – although none have produced the dramatic results seen in the Bullet. Using the results of almost 1,000 hours of observations with the Hubble Space Telescope to take images and the Subaru telescope in Hawaii to measures distances to galaxies in the images, astronomers have even managed to create a three-dimensional image of dark matter.

Meanwhile the Bullet Cluster stands as a reminder of the fact that the concept of dark matter continues to lead to convincing explanations, and thus is still worthy of being considered a valid theory. Although arguments continue as to the exact nature of this non-baryonic material, it is now clear that even modifying gravity does not eliminate our need to believe that dark matter – matter that we will never be able to see – is out there.

Why dark energy?

According to the latest estimate, the visible Universe – that is to say everything we can see: galaxies, stars, planets – accounts for only four per cent of the energy in the Universe. Twenty-three per cent takes the form of 'dark matter' and the remaining 73 per cent is put down to what is called 'dark energy'.

Up until this point in the history of the Universe, roughly seven billion years after the Big Bang, the expansion had been slowing down under the influence of gravity. Gravity was the only force capable of making a significant difference over astronomical distances, and it is attractive, attempting to pull matter back together. We might expect the strength of the gravitational force to determine the final fate of the Universe.

The Universe was expanding in the epoch that we are discussing, and it is expanding now. But will the expansion go on for ever, or will the galaxies turn back and rush together once more in a 'Big Crunch', many billions of years hence? Everything depends upon the average density of matter in the Universe, denoted by the Greek letter Ω (omega). If Ω is greater than 1, gravity remains dominant and in the fullness of time there would be a Big Crunch. If Ω were exactly one, then the expansion would continue to slow down but would never actually stop; this is known as a 'flat' Universe. If Ω were below this critical value, the expansion would slow down but would continue for ever. As we said when we considered inflation, the evidence we have seems to suggest that our Universe is flat, but observations of supernovae of a particular kind, known as type Ia, warn us that things may be more complicated.

We can see back to this critical epoch, roughly halfway between the Big Bang and the present day, by looking for these supernovae. Why are these particular explosions so special? It turns out they all peak at the same immense luminosity and they can therefore be used as standard candles, hence allowing us to measure distances. We compare how bright the blast should be with how bright it appears in the night sky, and the difference reveals how far away the supernova is. Supernovae that appear brighter than they should must be closer than expected.

Why should all these supernovae have the same intrinsic luminosity? A supernova of this kind is believed to be due to the total destruction of the white dwarf companion of an ordinary star. The small, dense dwarf pulls so much material away from its larger companion that eventually it becomes unstable, and there is a colossal thermonuclear explosion as the dwarf blows itself to pieces. As this explosion always occurs at roughly the same critical mass, the luminosity of the blast is the same in each case. While there are some differences which depend on factors such as the composition of the explosion's 'fuel', these can be adjusted for.

We have two ways of calculating the distances of the galaxies that hosted the supernovae; from the redshifts in their spectra, and from the peak luminosities of

▼ Very Large Array

A moveable array of 27 radio telescopes, each 25 metres in diameter, on the plains of New Mexico. The most powerful radio telescope on the planet, the Very Large Array had a starring role in the film *Contact*.

Invisible astronomy

We know that visible light accounts for only a tiny part of the electromagnetic spectrum, and only in comparatively modern times have we been able to build equipment enabling us to study what we may call 'invisible astronomy' from radio waves at one end of the spectrum to gamma rays at the other.

Some investigations can be carried out from the Earth's surface. Most people are familiar with the huge radio telescopes such as Jodrell Bank, which are really in the nature of large aerials. One certainly cannot look through a radio telescope!

Infrared astronomy can also be carried out from the surface of the Earth. However, many of the other regions of the electromagnetic spectrum are severely blocked by layers in the atmosphere of the Earth, and this means that we have to use space-based research methods, such as probes and satellites. This is true for almost the whole of X-ray astronomy, for instance, and there is an important satellite, the Chandra X-Ray Observatory launched in 1999, which has been immensely informative in this area.

If we had to depend only upon visible light we would be in the position of a pianist who is trying to play a nocturne upon a piano which lacks everything apart from its middle octave.

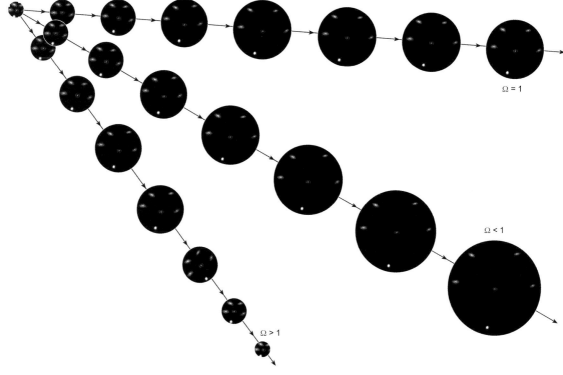

$\Omega = 1$

$\Omega < 1$

$\Omega > 1$

▲ **The fate of the Universe**
Depending on the total density of matter, the
Universe will either stop expanding at some point in
time and begin to shrink ($\Omega > 1$), expand forever ($\Omega
< 1$), or endlessly approach – but never quite reach
– a final size ($\Omega =1$).

the supernovae – but something is wrong. The supernovae look fainter than they
ought to do, and so it seems they are further away than anticipated. This was the last
thing astronomers were expecting. Only one explanation seems possible; the rate of
expansion of the Universe must be greater now than it had been before – the expansion
of the Universe must be accelerating instead of slowing down. It is the energy of this
acceleration that we call dark energy.

The fifth force

How can this be? Throughout the history of physics, only four forces were thought
necessary in order to explain all possible interactions between matter: electromagnetic

The supernova that shouldn't have been
The first evidence for the mysterious acceleration
of the Universe came from studies of type Ia
supernovae, which are believed to be 'standard
candles', always shining with the same luminosity
no matter where in the Universe they explode. In
the decade or so since that discovery, techniques
have become much more sophisticated, and
astronomers are now able to adjust for small

fluctuations in luminosity by studying such factors as
the time taken for the supernova to reach maximum
brightness and the speed of its decay. The principle
has remained sound, but one recent supernova may
change that. SN2006gy was decidedly odd. It was
extremely luminous, and calculations showed that
the mass of the material involved in the explosion
must exceed the Chandrasekhar mass

of 1.4 solar masses. This is the maximum mass
a white dwarf star can achieve, and so to see a
type Ia supernova with more matter casts doubt
on the accepted explanation for these objects. It
had a very odd light curve, and would have been
rejected by all of the cosmological surveys, but it is
nonetheless intriguing that we cannot yet explain its
extraordinary luminosity.

force (responsible for the attractive force between opposite charges), 'strong' nuclear (holding atomic nuclei together), 'weak' nuclear (causing radioactive decay) and gravity, the force of attraction which operates over the entire Universe. It is by far the weakest of the four forces, but it is dominant as far as astronomers are concerned simply because it is the only one that acts over large distances. (The electromagnetic force is also capable of long-distance interaction, but because matter is on average electrically neutral the forces cancel out.) Yet an accelerating universe required a fifth fundamental force that had not shown its effects earlier.

There are, however, theoretical speculations about a force that might fit the bill, most of which had been quickly discarded when first proposed. They lead us into the weird realm of vacuum forces and virtual particles. Quite naturally, we think of a vacuum being the complete absence of matter, but in the picture of the world we inherit from quantum physics this turns out to be an over-simplification. Any vacuum is really a seething, boiling mass of so-called 'virtual' particles, which always appear in pairs made up of a particle and an antiparticle. These virtual particles, carrying opposite charges, almost always last for only a tiny period, less than 10^{-43} seconds, before they must collide, annihilating each other. This process can be described as the vacuum 'borrowing' the energy it needs to create the particles and then giving it back by annihilating them before the rest of the Universe can notice. In their brief existence, however, these can have an effect on their surroundings – in the laboratory they have actually been seen to exert, in certain circumstances, a repulsive force. This could be just what we are looking for. Furthermore, the greater the volume of the vacuum involved, the greater the force, and so we expect the force to become greater as the Universe expands, exactly as observed.

Cosmic shear

Further evidence for the existence of dark energy has come from an unexpected source. By looking at the shapes of several hundred thousand galaxies, astronomers are able to measure the expansion of the Universe since the light was emitted from each galaxy. The method is known as 'cosmic shear' and relies on light being bent as it passes by matter. The most spectacular examples of this process are the Einstein rings, which are formed when light from a distant galaxy is so distorted by passing near a nearby system that it is spread out into a ring with the nearby system at the bullseye. Galaxies are also often seen distorted and stretched into arcs. Although these are dramatic examples, the image of every galaxy we see should be distorted in some way, and the magnitude of this distortion will reflect the amount of matter the light has had to pass to reach the

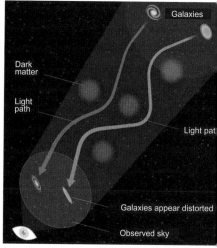

▲ **Cosmic shear**
Light from distant galaxies passes by clumps of dark matter. The mass bends the light, distorting the apparent shape of the galaxies.

▼ **Albert Einstein**
At the blackboard in Leiden, Holland, on December 6, 1923.

Einstein's greatest blunder?

Einstein's theory said that the Universe couldn't be static – it had to be expanding or contracting under gravity. But he was told by the astronomers he consulted that the Universe was static, and, perhaps surprisingly since his discoveries were based on his determination to follow the logic wherever it led him, added a 'magic' term, called 'the cosmological constant' – a repulsive force that counteracted gravity – to keep things in balance. In 1956, the year after Einstein's death, the physicist George Gamow wrote: 'Einstein remarked to me many years ago that the cosmic repulsion idea was the biggest blunder he had made in his entire life.', and the idea it was considered a blunder entered folklore. Except that such an antigravity force turns out to be needed to explain the accelerating expansion we see today, and could instead have been considered yet another theoretical triumph for Einstein.

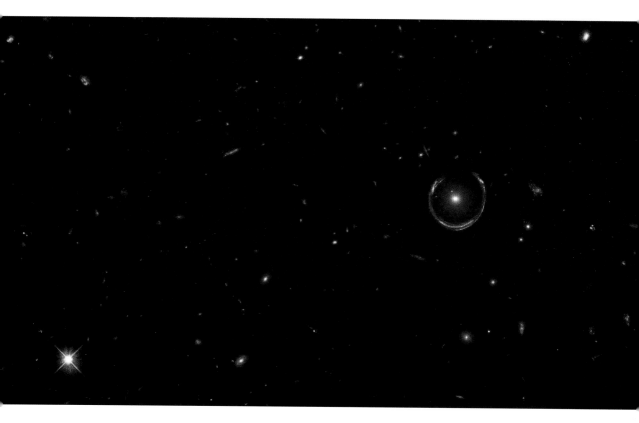

▲ Einstein ring

The light from a distant blue galaxy has been bent around a very powerful luminous red galaxy LRG 3 757 that lies between Earth and the blue galaxy to form an almost complete circle. This is known as an Einstein ring because Albert Einstein predicted this phenomena more than 70 years ago.

observer. For most galaxies, the effect will be weak and will manifest itself only by a slight tilt in the alignment of the galaxy on the sky. There is one problem, though. We only see the galaxy after this tilt has happened, and yet to measure the amount of matter passed, and hence the expansion, we need to compare the image we see with the image as it was when the galaxy emitted the light, before any distortion. For any particular galaxy, this is impossible, but with the huge numbers of galaxies placed at the disposal of astronomers by modern surveys it is possible to take a statistical average of many galaxies and extract information in this way. The results seem to be unequivocal – an accelerated expansion is necessary to account for the path light from galaxies takes in reaching us.

There is, however, a severe sting in the tail. Before the discovery of the acceleration of the Universe, particle physicists came up with a plethora of reasons why this effect – predicted by many of their theories – did not show up in our Universe. In fact, we are left with a situation in which it seems to be possible to explain why there is no repulsive force at all, or else why there may be an extremely large effect. Unfortunately, what has been observed is only a very small force (although added up across the whole Universe, of course, its effects are highly significant) and a major discrepancy remains. In fact, the

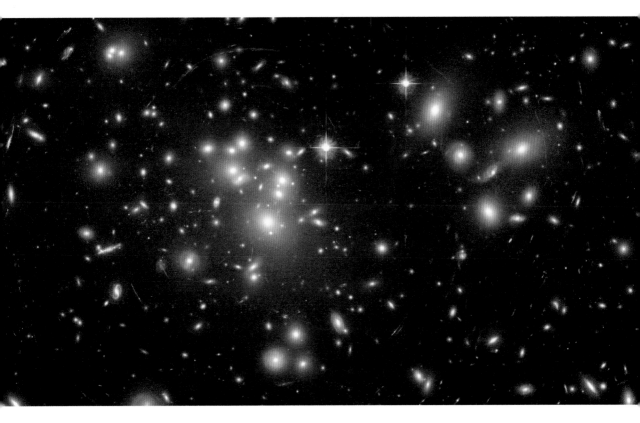

difference between the astronomical observations and the best theoretical model is somewhere around a factor of 10^{120}, a number that has to be the largest error between theory and experiment found anywhere in science at any point in history! However, this is the best explanation we have.

The situation may be even more complicated. We have assumed that the repulsive force is constant throughout all time, and yet we have no real reason for this assumption beyond the usual desire not to complicate things. (Occam's Razor, a principle often quoted by scientists, says that when all else is equal the simplest explanation is the correct one.) Some cosmologists believe that the strength of the force responsible for the acceleration would indeed vary over time. Large surveys of the sky, particularly with the Vera Rubin Observatory now under construction in Chile and ESA's upcoming Euclid satellite, will refine our cosmological measurements with the hope of determining whether dark energy does indeed change with time. Any such variation would be an extremely valuable clue as to what dark energy really is. For now, though, we are really in the dark.

▲ A giant gravitational lens

This amazing photograph from the Hubble Space Telescope stopped astronomers in their tracks when it was first revealed. It was an instant confirmation of Einstein's General Theory of Relativity, which had predicted that rays of light would be bent by the gravitational influence of a massive body. It is of the Abel Cluster 2218, a group of yellowish galaxies so massive that their combined mass forms a lens, producing a multitude of distorted, arc-like images of another group of galaxies much further away, seen as blueish. Sometimes multiple images of the same distant cluster are produced.

9 BILLION TO 9.2 BILLION YEARS A.B. (AFTER THE BANG)

CHAPTER 4 **Stars and Planets**

▲ The formation of our Solar System

Material left over from the formation of the Sun makes up a disk around the young star, whose powerful jets are clearly visible. As large bodies begin to form in the disk, collisions will be common, forming the orbiting planets.

▶▶ The Orion region

Halfway between the bright orange star at top left (Betelgeux) and bright blue star at lower left (Rigel) you can see the three stars of Orion's 'belt' with the stars of his 'sword' below, and the Orion Nebula behind the middle star of the sword. All of this surrounded by the vast C-shaped nebula known as Barnard's Loop. The dark silhouette of the eponymous Horsehead Nebula can be seen adjacent to the left-hand star of the belt. The Witch Head Nebula is the white form at bottom right.

▶ Dark clouds into stars

This is the Bok Globule B68, photographed in visual (left), and infrared (right) light. Clouds of dust and gas are a breeding ground for stars, and this strange dark cloud, seen here in silhouette against a bright background of distant stars, could be the precursor of many protostars. As can be seen from the right-hand image, shifting to longer wavelengths allows astronomers who study star formation, to peer through the clouds.

In previous chapters we saw the Universe lit up by the first stars, and the slow dance by which the galaxies assemble into their groups and clusters. At this point, nine billion years after the Big Bang, the Universe has started to settle into a state not so different from what surrounds us today. While for a long time, the most conspicuous parts of these galaxies would have been their bright centres, illuminated by the glow of material as it falls towards their central supermassive black holes. What is left once these quasars have used up their reservoir of gas and dust is a population of 'normal' galaxies, where the most prominent source of light is from second generation stars.

As we saw in the previous chapter, the first generation of stars ended their short lives in blazes of glory, with supernova explosions that spread heavy elements widely. It seems likely that the supernovae produced by these dramatic events also triggered the formation of new stars from neighbouring clouds of gas, as their shockwaves spread out through the galaxy. This second generation of stars are very different from their predecessors, precisely because they contain the heavier elements created in those early supernovae. This means that smaller stars can form for the first time, and this second generation of stars contains everything from tiny dwarfs to the most massive of giants.

As we will see shortly, these most massive stars will themselves explode as supernovae, and the debris from these explosions will be added to the mix. With each successive generation of stars born the amount of heavy elements increases. The Sun is thought to be a third generation star with a higher metal content than a second generation star which, in turn, has a higher metal content than the first generation of massive stars which were just made of hydrogen and helium. Around three billion years after the Big Bang there was a rapid increase in the rate of new stars being born. Since then the Universe has begun to decline into relative darkness, the end of the great cosmic dawn. Five billion years ago (around nine billion years after the Big Bang), in one spiral galaxy from the hundreds of billions in the observable Universe, a star we know as the Sun, was just beginning to form.

Evidence from meteorites, which include rare isotopes in samples of primordial material, suggests that the Sun may have formed from a clump of gas and dust which was influenced by just such a supernova.

Birth of a star

Star formation in galaxies is not an ordered or uniform process, and a degree of chance is involved. Whether a particular cloud of gas and dust will collapse depends to a great extent on the conditions in the surrounding material. Optical images of spiral galaxies like the Milky Way show a yellowish central bulge, surrounded by a disk of material in which spiral arms are embedded. These arms are much bluer than the rest of the disk, a colour which indicates the presence of young, hot, massive blue stars. These stars burn through their reserves of fuel rapidly, leading to short lifetimes of just tens of millions of years, compared to the many billions of years that their smaller redder siblings can manage. This means that whenever we see blue stars, we know we're looking at a place where star formation has happened recently, and we can conclude that star formation is concentrated in and around the spiral arms. Though while you're reading this, the Sun and Solar System find themselves in between two arms, on an interarm structure called the Orion Spur, which sits between the Perseus and Sagittarius spiral arms; their story likely starts inside one of the arms themselves.

More specifically, the Sun, like all stars, begins its life in a stellar nursery, known as a nebula, a reservoir of gas and dust. At the core of many of these objects are clusters of newly formed stars. It seems that stars are not often born alone, but rather that they form in groups called associations, or in denser clusters. The nearest of these stellar nurseries is the Orion Nebula, visible to the naked eye just below the three stars of the belt in the constellation of Orion. The nebula, which is around 1,350 light-years from Earth, is truly enormous – 24 light years across, and containing material with a mass equivalent to two thousand Suns. The nebula we observe is a small part of a much larger complex, which stretches across most of the constellation, and includes the famous dark Horsehead Nebula.

We essentially see the part illuminated by the Orion Nebula's cluster, a stellar beehive with a packed core. The central part's radius is not much more than half a light year across, and there are 670 stars packed into that space, which is not much bigger than the volume of our Solar System. (Remember, our nearest star, Proxima Centauri, is just over

▶ **The Southern Pinwheel Galaxy (M83)**

Galaxy M83, also known as the Southern Pinwheel Galaxy and NGC 5236, is a barred spiral galaxy approximately 15 million light-years away in the constellation borders of Hydra and Centaurus. It is thought to resemble our own Milky Way galaxy.

▶ **A Map of our Milky Way**

A map of the Milky Way as it might appear from outside our galaxy. It has two main spiral arms (Scutum-Centaurus and Perseus) attached to the ends of a thick central bar. Two minor arms (Norma and Sagittarius), where star formation occurs, lie between the major arms.

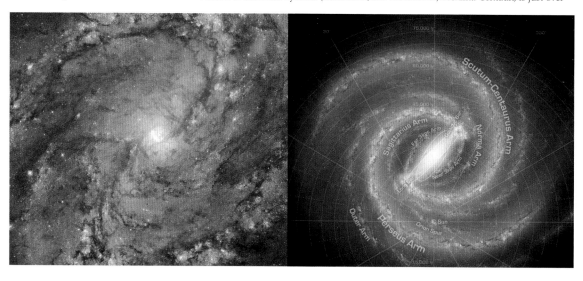

▲ Pillars of Creation

From perhaps the best known image from the Hubble Space Telescope, these pillars of interstellar gas and dust are a chrysalis for new stars as well as a work of art. Several of the newly formed stars can be seen emerging at the tips of spine-like features.

four light-years away). On the fringes of the Orion Nebula cluster, things are less tightly packed, but the density of two stars per cubic light-year still dwarfs anything seen in our present solar neighbourhood. These cramped conditions are what makes star formation possible, and our Sun's life must have started in just such a nursery.

In the same way that star formation does not occur uniformly across a galaxy, stars do not form uniformly throughout a nebula. Gravity is a weak force, after all, and it is surprisingly difficult to bring about the collapse of a cloud, which is what is necessary for a star to form. In normal circumstances, the random motion of the particles in the gas, stirred by the movement of the cloud in the disk of the Milky Way, and by the influence of pervasive magnetic fields that thread the region, would prevent a collapse.

In denser regions the odds are better because here, the hot, giant stars of the cluster can help. The radiation pressure from these stars pushes the dust and gas away, forming the sculpted bubbles you can see in the Orion Nebula image. It is on the dense edges of these cavities that stars will form. In these regions the density will be between 100 and 10,000 atoms per cubic centimetre. This is the equivalent of the vacuum that would surround you if you stood on the Moon, but it is impressive compared to the density of interstellar space, which has only about one atom per cubic centimetre.

The increased density makes collapse under gravity much easier, so clumps of yet denser material will start to form. These will be cooler than their surroundings because the dust that they contain shields them from the ultraviolet radiation that pervades the nebula. In the early stages the clump will cool due to radiation from the hydrogen atoms that make up the majority of the gas, but as the temperature drops the more efficient carbon and oxygen atoms can take over, followed by radiation from molecules such as carbon monoxide. At this stage, the dust can also radiate heat in the form of infrared radiation, which cools the clump further. As the temperature drops to no more than 10 Kelvin (K) (-263 degrees Celsius), the gas and dust have slowed enough that gravity is, at last, the most important force and the collapse can proceed.

As the material that will make up the new Sun collapses, it will begin to spin, obeying the law of conservation of angular momentum. As something contracts, it spins faster and faster, just like an ice skater spinning on the spot who can speed up by pulling their arms in. By this point, the speed with which the forming star is spinning will mean that some material will be flung away, and so not all of the cloud will end up in the final star. As this material is lost from the cloud, it carries away angular momentum. Imagine our skater holding onto a basketball; if they throw the ball away their spin will slightly slow.

The collapsing clump is now 0.2 light-years across. Within it, a denser pre-stellar core is being formed. At first it accounts for only a tiny fraction of the total cloud mass, weighing maybe only a hundred thousandth of the mass it will reach at maturity, but this is enough for gravity to keep the collapse going. As its density rises, the cloud becomes increasingly opaque to infrared radiation, which is trapped. As a result, the core begins to heat up.

This heating causes an outward radiation pressure from the core, which soon stops more material from falling onto it. The core itself is still collapsing under its own gravity, however, and this collapse heats it further. When it reaches a temperature of a few thousand degrees, it is hot enough for the hydrogen molecules within it to break into their separate atoms. At this point the innermost region is just a few astronomical units across; we have reached the scale of a solar system. The increased density causes further collapse, and when the temperature climbs to over 10,000 K the electrons escape from their nuclei, and the hydrogen becomes an ionised plasma. The formation of a star is now all but inevitable. The free electrons in the ionised plasma trap more of the radiation in

How bright is a star?

Magnitude is a measure of a star's apparent brilliance. The scale works rather in the manner of a golfer's handicap with the most brilliant performers having the lowest values. Thus a star of magnitude 1 is brighter than one of magnitude 2, which is brighter than one of magnitude 3 and so on. On a clear night, an average person can see (from a dark site) down to magnitude 6 with the naked eye, while modern equipment can see down to magnitude 30.

At the other end of the scale, the brightest star, Sirius, is as at magnitude of –1.5, while the brightest planet, Venus, can surpass –4. The Sun has a magnitude of –26.7.

this denser, protostellar core, leaving this hot object surrounded by cold gas and dust.

The evolution of a newly collapsing cloud into a protostellar core takes just a million years or so, fast by astronomical standards, but it is hard to see. The presence of the dust which enabled the collapse in the first place makes it difficult to get a proper look at these stages of star formation, though observations with radio and submillimetre telescopes can help. These instruments can detect the material which surrounds the core. At a temperature of 300 K (the same as a hot day in a British summer!), this material is a molecular soup, though the density is low enough that the molecules are, on average, a thousand kilometres apart. Observing the rich diversity of this chemistry can tell us a lot about the future life of the protostar and the planets that will soon form around it.

Chemicals for life

Images of a nebula like Orion reveal a subtle and varied pallet of colours, which carry information about its composition and state; in our picture the deep red is caused by the ionisation of hydrogen. Tuning into this wavelength allows us to measure the full extent of the nebula. The bright blues, in contrast, are reflection nebulae, with the light from the bright young stars at the heart of the stellar nursery reflecting off grains of dust.

These particles, each a tenth of the size of a sand grain, are scattered throughout the nebula. Together, they make up only one per cent of the mass of the nebula as a whole, but it's the presence of these particles that makes many parts of these star-forming regions opaque to optical light, which is why we turn to infrared and particularly to sub-millimetre wavelengths to study them. Spectroscopy reveals that the gas consists not only of hydrogen, but also of oxygen and many simple molecules, such as carbon monoxide and water, alongside more complex species such as the alcohols, methanol and ethanol (CH_3OH and C_2H_5OH respectively).

Molecules of this complexity normally form when smaller molecules collide, but when the material is cool and so rarefied, clearly collisions can't happen very often. So how does such a rich chemistry arise around the protostellar core? Again, the answer lies in the presence of dust. Gas molecules settle onto the surface of the dust grains, and this can bring the molecules into contact with each other. Hydrogen atoms in such circumstances can only be loosely held, and they naturally move around the surface as a result. The situation is rather like that in a catalytic converter, which helps remove pollutants from a car's exhaust by providing a surface on which harmful molecules can react. In the nebula, even the formation of the simplest molecule of all, molecular hydrogen – H_2 – from individual hydrogen atoms requires such a boost; without the dust, the nebula would be a very different place.

In all, more than two hundred different types of molecule have been detected in giant clouds of gas and dust like the Orion Nebula. Some of these molecules are very complex indeed; even the amino acid glycine (NH_2–CH_2–$COOH$) may have been detected. On Earth, such amino acids are the base units of proteins, and hence they are the building blocks for life. Amongst them are the molecules which make up the genetic material which is shared by all life on Earth.

The presence of amino acids does not necessarily mean life. However, the fact that such complex molecules which do play an essential role in life exist in star forming nebulae suggests that they may be present in the material which will shortly be assembled into planets. If so, they could conceivably jump-start the more complex chemistry of life once the planets have formed.

The fact that much of the chemistry that takes place in the nebula is carbon based is particularly interesting. Carbon can form up to four stable chemical bonds at once, with a wide variety of different molecules, an ability that is responsible for the extraordinary richness of carbon chemistry and the diversity of carbon-based molecules that we see. No other atom is as versatile; silicon, the atom directly below carbon on the periodic table has similar properties, but cannot easily form molecules made from long chains of atoms in the same way as carbon. So, as yet, silicon-based life remains in the province of science fiction.

Carbon's ability to form four bonds at once also introduces a property known as chirality. Picture a carbon atom bonded to four molecules, each of which are different. There are two different ways you can arrange things, each a mirror image of the other, and we call these two possibilities the left-handed and right-handed forms. Both have the same chemical formula, and consist of the same five components, but because of their different arrangements they may have slightly different chemical and physical properties.

Take the chemical carvone, for example, found in many essential oils (though not – yet – in the Orion nebula). One form of carvone, the right-handed one, smells of spearmint and is used in chewing gum, while its equivalent left-handed form smells of caraway seeds. Both forms have the chemical formula $C_{10}H_{14}O$ but your nose can tell the difference.

This turns out to be a (very) specific example of a general principle. Simple chemical processes should produce equal numbers of left- and right-handed molecules, and treat them equally. For complicated chemistry, including that taking place in your nose when you smell something, it turns out the difference between left- and right-handed forms does matter.

In fact, we can go further. Life, on Earth at least, has a preference for left-handed molecules, basing its chemistry on these exclusively. Why this should be so is still a matter of some debate, but it is at least possible that the explanation might go right back to the time of the Solar System's formation, and to conditions within the protostellar nebula itself. It turns out that the way that light scatters off dust in the middle of a nebula can – because of a property known as circular polarization – selectively destroy right-handed but not left-handed molecules (or, depending on the situation, vice versa).

So if the molecules which started the long process of life's chemistry on Earth had their own start not on the surface of our newly formed planet, but in the protoplanetary disk from which it formed, it may have been natural for one form or other to be favoured. A basic fact about life on Earth may betray our cosmic origins – and if we ever find life elsewhere in the Solar System we might expect it also to have left-handed chemistry. Of course, just because a complex chemistry exists in the raw material from which planets form, it doesn't mean that those molecules will make it unscathed onto the surface of the forming world.

To understand how likely this is, we need to return to our forming protostar. The core, now the same size as the present day Sun but only a thousandth of its mass, about the same as Jupiter, is still opaque. It is, however, growing rapidly as material surrounding the central core accretes onto it, gradually increasing its density. Over the course of ten thousand years or so, its mass will get a lot bigger, increasing at least tenfold.

How big the star will eventually become depends on the amount of material that is available, but we do know a major change will take place when the core has acquired a

▶ **Orion Nebula**
The vast regions of gas and dust seen in red in this image from the Hubble Space Telescope are where thousands of new stars are being born. More than 3,000 stars in various stages of development are contained in the field. The bright central region is the home of the four most massive stars in the nebula, called the Trapezium because of their geometrical pattern. Ultraviolet light from these stars creates a cavity in the nebula and influences the growth of hundreds of smaller stars. Within the Trapezium are stars encircled by planet-forming "protoplanetary disks", but they are too small to be seen in this picture.

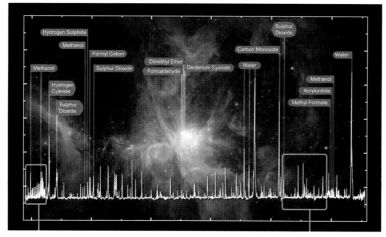

◀ **What's in a nebula?**

The HIFI spectrometer flew onboard the Herschel
Space Observatory. Its job: to identify water and
organic molecules in the Orion Nebula. The far-
infrared spectrum it recorded is superimposed on an
image of Orion taken by the Spitzer Space Telescope.
The presence of organic molecules in the spectrum
could eventually provide clues as to the chances of
life emerging if and when planetary systems form
around stars in the region

mass of about half that of the present day Sun. At this point, the temperature at the centre
of the core will be high enough, at about a million Kelvin, for the first stages of nuclear
fusion to start. As we will see, a race against time has begun; this fusion will eventually
stop the accretion of new material onto the star entirely, but for now it continues to grow.

Once the core has reached ninety per cent of the mass of the Sun, the outward pressure
from radiation generated by the nuclear reactions at its centre prevents further gravitational
collapse. Material can still make it to the star, though, because its path can be shaped by the
strong magnetic field generated by the star. The process of accretion is now far from smooth.
As clumps of material are drawn in, their accretion causes the star to undergo many violent
outbursts. The magnetic field can also produce powerful jets, which accelerate material to
great speeds, carrying it away from the star's poles. The situation is unstable, but despite this
unsettled period, the star is able to grow to be the mass of the Sun.

It is now known as a T Tauri star, a phase named after the first star to be observed at
this point of its evolution. It is still shrouded in gas and dust, most of which is contained
within a flattened disk which stretches over hundreds of astronomical units, many times
larger than the present-day Solar System. In this disk is the material from which planets
will form, but let us concentrate first on the life of the star at the system's centre. Over the
course of the next hundred million years the temperature in the centre of the protostar
will rise, and eventually it will exceed 10 million Kelvin, the critical point at which it is
hot enough for hydrogen nuclei in the core to fuse reliably together. If there is a single
moment when a star becomes a star, this is it; the outward radiation pressure from
nuclear fusion can now balance the inward gravitational pull, creating the equilibrium

◄ **T Tauri**

The yellowish star near centre in this telescopic
view is T Tauri, the prototype of a class of
variable stars. Nearby it is a dusty yellow
cosmic cloud known as Hind's Variable Nebula
(NGC 1555). Over 400 light-years away, at
the edge of a molecular cloud, both star and
nebula vary significantly in brightness but not
necessarily at the same time. T Tauri stars are
now generally recognized as young, less than a
few million years old, Sun-like stars still in the
early stages of formation.

which will sustain the star throughout the majority of its life. It will remain in this new,
relatively stable state for a period many times longer than it took to form.

The newly ignited star now lies on what is known as the 'main sequence', a feature
which reveals itself when you plot the colour and brightness of a population of stars
against each other. In such a graph, known as the Hertzsprung-Russell (HR) diagram,
the majority of stars lie somewhere along a snaking line – the Main Sequence – which
reaches from hot blue stars to faint red dwarfs. All the stars belonging to this sequence are
happily generating energy by fusing hydrogen at their cores to form helium, and how long
they will stay in this state depends on their mass.

A star like our Sun, living pretty much in the middle of the main sequence, will last
around 10 billion years in this state, over 80% of its total lifetime. Smaller stars, less
massive than the Sun, will last even longer, while hot, blue, massive stars will use up
their fuel quickly, and have shorter main sequence lifetimes. The explanation for this
variation is that the process of nuclear fusion is regulated by a natural thermostat. If the
star contracts under the influence of gravity, then the temperature at the core will rise,
reactions will occur more frequently, more energy will be produced increasing the outward
pressure, and the star will inflate, reducing the temperature at the core. The position of
this equilibrium, and the temperature of the core, depends only on the mass of the star, so
the word 'sequence' is rather misleading. Any star remains in the same position on the HR
diagram throughout its main sequence lifetime; to be a member of the main sequence is
to have found your position and to stick to it.

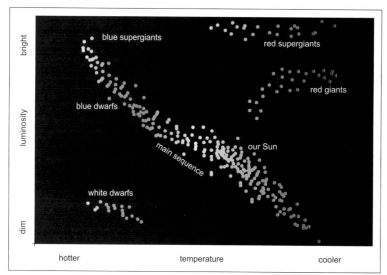

◄ Hertzsprung–Russell diagram
A Hertzsprung–Russell diagram is a way to illustrate the differences between types of stars, plotting luminosity as a function of temperature (or colour, since hot stars are blue and cool stars red, the two are equivalent). Most stars lie on what is called the 'Main Sequence', which runs from top left (hot, massive, luminous blue stars) to bottom right (cold, small, faint and red stars). The term *sequence* should not be taken to imply that stars move along it as they age. Toward the end of their lives, most stars leave the Main Sequence, moving off to the right and becoming giants. The most massive stars leave first, then the intermediates like our Sun, which will spend a total of eight billion years on the Main Sequence, and finally the smallest dwarfs. An older cluster will lack the blue end of its Main Sequence, and that can be a valuable way of determining the age of clusters in distant galaxies.

Formation of the planets

While we have been distracted by the forming star, out in the disk of material surrounding the protostar, planets have taken shape. The disk holds only a tiny fraction of the system's mass, and much of this will be lost as it evolves. The gas in the disk is either rapidly accreted onto the star or quickly blown away by the violent eruptions associated with the star's youth. Meanwhile, some of the dust will clump together to begin the process of planet formation, some will be ejected from the system via random gravitational interactions, and some will fall into the star. Studying this material is hard; it is often near-invisible in the optical region of the spectrum, reflecting very little of the light emitted by the protostar. The glare of the protostar also makes peering into the regions where these processes are happening particularly difficult. A dust disk will, however, betray its presence via an excess of emission in the infrared and submillimetre regions of the spectrum; the wavelength we use to detect the particles will depend on the size of the material which makes up the disk. We need to use shorter wavelengths to detect smaller material. This effect has been exploited by astronomers, who, reasoning that aliens mining asteroids would inevitably generate dust, looked at nearby debris disks to see if there was an otherwise unexplained excess of small particles. No evidence of alien mining was found.

We know from what we can see in nearby systems that the gas in the disk will be gone by the time the star reaches the end of the T Tauri phase, and the dust will disappear well before the star joins the Main Sequence. This means planets must form quickly; in under 10 million years for giant planets, which need to grow to their final size while there is gas left for them to accrete – and 100 million years for smaller, rocky or icy planets.

Early on, the disk will flatten due to the spin of the system – rather like a ball of dough being spun into a flat pizza. The shape of the disk is defined by both the behaviour of the protostar and any external radiation from nearby stars or other astrophysical objects. The protoplanetary disk starts roughly 5 million miles from the central T Tauri star; any closer than this and the star's radiation will prevent a stable orbit. From the inner edge of the disk, hot, ionised material is funnelled along the magnetic field, eventually looping down to be accreted onto the central star near its poles.

Close to the star, the disk of mixed gas and dust is narrow, but then it fans out further away from the harsh radiation of the forming star. The top and bottom of the disk are exposed to external radiation from the interstellar environment, which helps shape it, preventing it from bulging like a rugby ball.

Toward the edge of the disk, getting further from the protostar, the temperature drops. Only at the inner edge is it high enough to ionise the material, and so only there will the magnetic field be so important. Further out, the composition of the solid material in the disk itself will change, because each element or molecule will condense from a gas to a solid at a different temperature.

As we work our way from the protostar outwards through the disk, aluminium is first to change from a gas to a solid. This process happens at a temperature of about 1,500 Kelvin. Next comes iron, a little further out, and then, as the temperature decreases, magnesium

▼ **Protoplanetary disk**

Cross-section of a protoplanetary disk showing the stratified gas structure shaped by external radiation from the protostar and the temperature gradient away from the forming star resulting in build up of material at the ice lines. Observations over different wavelength ranges can look at different regions of the disk where the planets will eventually form.

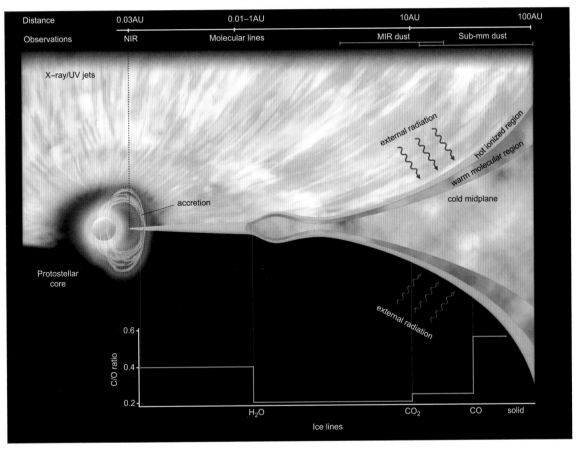

silicates, familiar to us as beach sand. Each of the dust grains produced by this process is just a fraction of a millimetre across, but from these tiny seeds vast planets will soon grow.

We are now about as far from our protostar as the Earth is from the Sun. Continuing outwards, we reach the important point where the temperature is 0 degrees Celsius (273 K). From here on out, water will freeze. This is important because water is so abundant; it is the third most common molecule in the Universe, and it is plentiful in the kind of environments where stars and planets form.

As we cross each of these, more material will freeze out, and so as we move outwards in the disk there is more solid material available with which to form planets.

The newly solidified material joins the dust which already exists throughout the disk. A typical dust grain is made of carbon or silicon, and will have a diameter about a tenth of that of a sand grain. As different materials freeze out, they may freeze onto existing dust grains; water, for example, is eight times more likely to freeze onto a surface than to condense out spontaneously on its own. So dust grains here in the outer disk are surrounded by a mantle of ice crystals.

The ice forms a thin coating, and plays an important role in what happens when two dust grains collide. The icy mantles soften the blow of such a collision, meaning that the colliding dust particles are much more likely to stick together afterwards. In some cases, where the relative speeds of the two particles were slow to begin with, the icy mantles themselves can act to stick the dust grains together.

Larger clumps are soon produced, and as this process accelerates, pebbles, millimetres or even centimetres across, collide and combine to build up larger and larger rocks. Evidence from meteorites suggests that this process starts soon after the disk settles; samples which date from a time just one million years after the formation of the disk that produced our Solar System contain millimetre-sized clumps of material.

It is difficult to determine how these small rocks then build up to form ever larger boulders, yet they must eventually produce planetary embryos which are hundreds or even thousands of kilometres across. The process of colliding rocks and getting them to stick together rather than breaking each other apart is a delicate one, and we don't yet understand it. What we do know is that the process will proceed differently depending on the location within the disk. Out beyond the water ice line there is much more material available in solid form, so the planetary embryos formed out there will have an icy crust around their rocky core. This added mass turns out to be a significant contribution, as adding the ice means that large bodies can be assembled rapidly.

Until the bodies are larger than about a metre, they move along with the gas as it orbits the star. As they grow, though, they begin to move independently – we say they have 'decoupled' from the gas. But this means that as they are no longer travelling with the gas, so they will feel a force based on their relative motion and the net result is a change in the planet's orbit. This form of migration – known as type I migration – will allow the protoplanet to keep growing rapidly as it will continually encounter new material. This is the solution to the problem of how planets can grow efficiently, solving the problem of how to assemble such large worlds in the limited time available before the disk dissipates. The effect will be larger for more massive protoplanets, and so these will tend to grow faster than their smaller counterparts.

As the protoplanet increases in size, it extends its gravitational reach further and further, growing still more. We have entered a runaway phase as our protoplanets grow exponentially.

If a planetary core can reach a mass several times larger than Earth, its gravitational pull will be strong enough to draw in and hold onto gas from the disk, so these worlds will quickly accrete large atmospheres made mostly of hydrogen and helium. Giant planets are more likely to form here in the outer disk, beyond the water ice line, where this process occurs most rapidly.

It is, however, a self-limiting process. Eventually, the planet is large enough that the drag from the gas is insignificant, and so its orbit will stabilize. It quickly accretes all the material that shares its orbit, carving out a gap in the disk. Remarkably, we can now see this process in action by seeking out stars with disks in nearby star forming regions. High on a desert plateau in northern Chile, an array of telescopes called ALMA does just that. ALMA uses short wavelength radio (mm or sub-mm) to see through the dusty gloom, revealing disks with exactly the kind of gaps we expect when planet formation is underway.

In some of the disks ALMA has seen multiple gaps, suggesting that several planets are forming at once in these systems. Some gaps may also be produced in a different way; if a planet is more massive than Saturn is, then its gravitational influence can also produce a pattern of smaller gaps either side of the main one.

In between the gaps, stable regions of higher than normal density provide a reservoir of dust within which smaller, rocky planets begin to form. Normally, such bodies would migrate inwards by the same process that affected the movements of the larger planets, and eventually they would fall into the star. But they cannot cross the gaps opened up by the more distant giant planets. The shepherding of the dust particles by the gravity of more distant, larger planets creates a safe haven where smaller worlds, like the Earth, can grow.

There are other processes at work in the disk too, shaping our planetary system. In particular, the giant planets may also move rapidly through the disk, in a process called type 2 migration. The complexity of these processes of dust and gas dynamics means that every system will form differently, and we should expect to see an incredible diversity of worlds around us in the galaxy.

Exoplanets discoveries

While astronomers have been explaining the origin and composition of our Sun's family of planets for hundreds of years, the story we told just now has only come together in the last thirty years or so. The idea of planets moving about while they are forming was only seriously considered once planets in other systems – exoplanets – had actually been found.

In 1995, astronomers studying the nearby star 51 Pegasi found that it appeared to be wobbling back and forth, movement that revealed itself through a regular pattern of Doppler shifts in its spectrum. The observations suggest that the star is in orbit around a position just slightly offset from its centre. By measuring the size of the wobble it was possible to get an estimate of the mass of the perturbing object, which turned to be less than half than the mass of Jupiter, too small to be a star. They had found an exoplanet!

The discovery was puzzling. The new planet completed an orbit every 4.23 days. That placed this giant world, which we would have expected to find in the icy reaches of its outer solar system, eight times closer to its star than Mercury is to the Sun. To find a giant planet so close to its star that its year was measured in just a few days was completely unexpected, calling into question everything we'd assumed about planetary systems.

One such planet might have been an exception, the result of some gloriously unlikely freak accident of planet formation. But 'hot Jupiters' like the planet perturbing 51 Peg were found to be common. More than 400 are known to date, and they account for around 10% of known exoplanets. This overstates their actual abundance – it is easier to find a large planet close to its star, where it will induce significant wobbles, than to pick out the signal of a puny Earth sized world. Correcting for these biases, hot Jupiters seem to account for about 1% of all worlds. That doesn't sound like much, but it could easily mean there are a billion hot Jupiters in the Milky Way alone!

The method used to find the planet around 51 Pegasi, known as the 'radial velocity method', requires enough telescope time to pay close attention to a star of interest over many nights. Most exoplanets known today have been found instead by searching for transits, a method which allows us to survey many stars at once. This technique relies on the fact that, if a planet's orbit happens to be aligned such that – as seen from Earth – it passes in front of the star, the star will appear to fade just slightly. A similar effect is seen from Earth when Mercury or Venus transit in front of the Sun. Thanks to the complexities

◀ **The Atacama Large Millimetre/submillimetre Array (ALMA)**

ALMA is a 66 dish array of telescopes that observes the Universe at wavelengths that lie in between those of radio and infrared radiation. Millimetre and submillimetre radiation is easily absorbed by moisture in the air, and the location 5,000 metres (16,500 feet) high up in the Chilean Atacama Desert was carefully chosen for its dry climate.

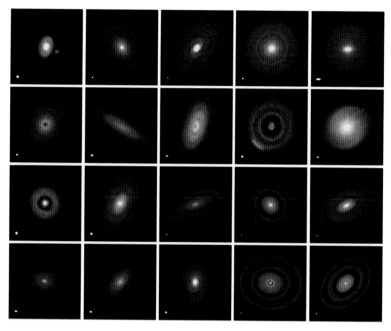

◀ **Protoplanetary disks**
ALMA made the first large-scale survey of protoplanetary disks around nearby stars. This incredible image shows 20 nearby systems where dust and gas are forming new planets. It is possible to make out gaps in the disks, which seem to demarcate the inner and outer portions of the disk.

of planetary geometry, such events are rare, but throughout the galaxy there must be worlds which happen to be positioned such that the planets of our Solar System cross in front of the Sun with every orbit, causing a regular series of blinks. Indeed, the nearby dwarf Teegarden's Star seems to have just such a system; if there are astronomers on these planets, what could they learn about our Solar System?

They could measure the radius of each planet relative to the size of the Sun; the amount of light blocked by the planet corresponds to the ratio of the two areas. By timing the period between successive transits, they would get the orbital period and – thanks to Kepler's laws – the distance of the planet from the star. (You can see why large, close-in planets would be easier to detect via this method than smaller planets which are further from their stars). Applying these techniques to the known hot Jupiters tells us that they span a range in size that starts from 100 Earth masses (about a third of the mass of Jupiter) and which goes all the way up to 13 times Jupiter's mass. This higher figure is most likely a fundamental limit; any more massive than this, and the pressure and temperature at the core will be high enough to enable deuterium fusion, at which point the body is at least a brown dwarf and perhaps a full-blown star.

This range of masses makes it clear we're not dealing with planets like our own Mercury, the only world that lies close enough to the Sun that it completes an orbit in less than 100 days, a range that includes three-quarters of known exoplanets. How did these strange worlds come to be?

One part of the answer is that it is sometimes easier to form planets efficiently. Perhaps unsurprisingly, there's a correlation between the metallicity of the star – how much material there is in forms other than hydrogen and helium – and the likelihood that giant planets would exist. The more material there is from which to form planets, the more planets form! But, though this relation sounds like nothing more than common sense, it doesn't seem to hold for planets smaller than Neptune. This suggests that there is a fundamental threshold that needs to be reached before a giant planet can form.

The necessary conditions must be reached more often in the outer protoplanetary disk, where we have already seen that planet formation can proceed rapidly. To reconcile these hot Jupiters with a theory of planet formation that says that such large worlds must form out beyond the ice line, the idea of planets migrating through the disk as they form becomes essential. To get to be a hot Jupiter, these worlds must have travelled a significant distance, ploughing through the inner disk, gathering up and expelling material through their gravitational pull as they did so. This is probably bad news for any small rocky planets with ambitions to remain in a nice sedate, stable orbit, but it also raises a fundamental question – how do you stop such a planet from falling into the star?

The answer is that most of the time you probably don't. In many systems, planets will form, migrate through the disk and disappear into fiery oblivion. But in some cases, the inner gap opened up by the heat of the star at the inner edge of the disk may act as a barrier, with the build up of material at the disk's edge slowing the migration enough for the newly hot Jupiter to settle into the orbits we see today.

What of the 99% of planets which are not hot Jupiters? The most prolific planet hunter was the Kepler Space Telescope, designed specifically to detect transits. Following launch in 2009, it spent three years staring at a single patch of the night sky, chosen to be rich in stars and yet devoid of any particularly bright examples, on the border between the constellations of Cygnus and Lyra in the northern part of the sky. Kepler measured the

▼ **Surveying exoplanets**

The Kepler spacecraft's mission was designed to look at a region of the Milky Way and determine the fraction of the hundreds of billions of stars in our Galaxy that might host exoplanets. Kepler discovered thousands of transiting exoplanets, getting a measure of their radius relative to their stars. This diagram charts the number of planets it discovered orbits of less than 100 days, showing there are a large number of worlds quite unlike those in our Solar System.

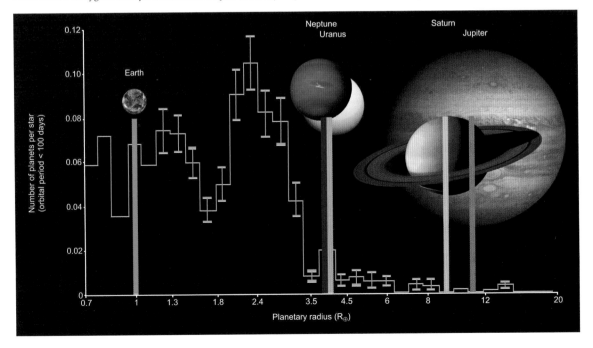

brightness of each of 150,000 stars every thirty minutes, and that of a smaller number of selected systems every minute. Its intended quarry was planets where life like our own might, perhaps, exist – terrestrial worlds around Sun-like stars, preferably at the right distance from the star that the temperature would allow for liquid water. What it found was much more interesting.

It turns out that the most common type of planet in the Milky Way is one that does not exist in our own Solar System, with a radius between that of the Earth and that of a planet like Neptune, four times larger than Earth. Members of this new class of planets are known as super-Earths, if their density suggests that they are rocky, or mini-Neptunes if a lower density indicates a gaseous nature. Ideally, we'd know not only the radius, which can be determined by the transit method, but also the mass, for which we need a detection via radial velocity.

Unfortunately, most of the stars studied by Kepler are too far away or too faint to enable useful radial velocity measurements to be made. A new NASA satellite, TESS, is searching for planets around nearby stars, partly to correct this problem. So far, what we know is that mini-Neptunes and super-Earths are both so common that it is odd that neither exists in our own planetary family. If we take all the planets found by Kepler on orbits less than 100 days, adjust for the fact that larger, close in planets are easier to see, over half of the detected planets still fall into this category.

Actually, this exercise reveals another new truth about the planetary population. There's a distinct absence of planets with sizes right in the middle of the gap between Earth and Neptune; at 1.7 Earth radii in particular there are few planets, a feature that has become known as the radius valley. It seems that planets really are either a super-Earth or a mini-Neptune, lying on one side of the valley or the other. The explanation comes from the time when planets are still forming. As the star heats the disk, only the most massive planets can hang on to their atmospheres. If the forming planet is not massive enough, its atmosphere is lost and it will end up as a super-Earth, while the more massive worlds can hang onto the gas and remain mini-Neptunes.

The details of these processes, and how they might have influenced our own world's formation, are still being worked out, but the existence of this fundamental division between planetary types which we were completely oblivious to until a few years ago gives you an idea as to how fast things are changing. Careful observations of planets on both sides of the radius valley, along with their stars, will be a big part of astrophysics in the next decade or so.

A third big surprise from our exoplanet discoveries is that a third of known worlds have orbits which are distinctly eccentric. It's true that the Earth's orbit isn't precisely a circle, but it's pretty close – its eccentricity is 0.016, so the difference in length between the longest and shortest axes is not much more than one per cent. A third of exoplanets have eccentricities greater than 0.1, an order of magnitude larger. This fact is a clue that life in a forming solar system may be even more complicated than we had suspected until now.

We've talked so far about planets moving through the disk, usually by interacting with the material in the disk itself. These large eccentricities were caused by more dramatic interactions between planets – so the fact that the planets of our Solar System have largely circular orbits tells us it must have been an unusually calm place when they were forming.

Even more extreme interactions are possible. When two large bodies come close together, their spheres of gravitational influence overlap. Anything caught between them

▼ Brian observing the transit of Venus in 2004
When a planet crosses the face of a star, in this case our own Sun, there is a slight dimming in the brightness of that star measured from Earth. This principle is used to identify exoplanets orbiting other stars.

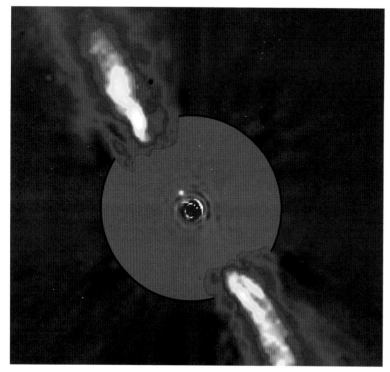

◄ **Beta Pictoris**

The star Beta Pictoris is a well-known example of a star with a surrounding "debris" disk. This star was the first to have an exoplanet imaged directly. Infrared observations made from the European Southern Observatory (ESO) in Chile revealed a planet orbiting the star (whose bright presence has been artificially removed by the large disk to increase the contrast needed to reveal the planet).

will become dynamically excited – in other words, once stable orbits will be disrupted, and material can be expelled from the system. It's even possible for one of the planets to be expelled, spinning off into space.

Some interstellar wanderers have even been detected, thanks to a technique originally developed for looking at distant galaxies in the form of gravitational lensing. If a planet passes between us and a distant star, it acts as an otherwise undetected lens, bending and amplifying the star's light and revealing its presence. Distant planets – even on the other side of the Milky Way – can be found by this technique, known as microlensing as the changes are small. What's more, the amount the star brightens tells us about the mass of the planet and several of those detected have masses comparable to Earth.

It's clear we can learn a lot from these indirect methods of planet detection. Indeed, for most systems we have no choice but to rely on these techniques, where we learn about the planets by studying stars. It is, though, undoubtedly true that direct imaging – seeing the planets themselves – can tell us much more. The problem isn't so much that the planets are intrinsically faint – especially after formation, when they are still being heated by their gravitational contraction, they will shine brightly – but that the glare of light from the star makes them very difficult to detect.

The solution is to use an instrument called a coronagraph to block the light from the star. By placing an obstruction in the field of view, and through some very careful use of image processing algorithms, planets can be revealed. It is easier to see companions which are further from the star, and direct imaging has been used to find worlds tens or even hundreds of AUs from their star. There are only fifty or so planets which have been directly imaged, each of them extremely precious. They tend to be young, and massive – many are nearly large enough to be stars.

Comparing each of our exoplanet systems to our own raises many questions. Why are the worlds of the Solar System on nearly circular orbits? Why don't we have a super-Earth? Why did Jupiter remain where it was, rather than ploughing through the inner Solar System and becoming a hot Jupiter? It is time to look at the origins of our own Solar System in the light of these new discoveries, learning more about our own home now we've looked outwards at the stars.

Formation of our Solar System

Our Solar System may not have suffered the chaos of Jupiter ploughing inward toward the Sun, but this is not to say that everything was stable from the beginning. It's just that the real drama came a little later, five to ten million years after it started forming. At around this time, Jupiter did go for a wander, via a migration that explains some of the more unusual features of our Solar System.

This idea, which has become broadly accepted in planetary science in recent years, is known as the Grand Tack after a manoeuvre in sailing where a ship rapidly changes direction. This name suggests sudden change – so what happened?

It's believed that Jupiter initially formed near the water ice line, about 3.5 astronomical units (AU) from the Sun. It would, like many other giant planets, have begun to migrate inwards. This moved the giant planet nearer to where Mars orbits today, between 1.5 and 2 AU from the Sun, and it swept up and disrupted the planetesimals in the inner Solar System. This explains the first of the Solar System's mysteries, the rather puny status of Mars, which is not as large as might be expected from its position.

So Jupiter grew fat by stealing from Mars. But why did it not continue to migrate inwards? The culprit, according to the Grand Tack theory, is its sibling Saturn. Saturn, then as now, was less massive than Jupiter, and so it was more affected by drag, moving more quickly towards the Sun. This meant it caught up with Jupiter, and the two settled into a resonant orbit. For every three orbits Jupiter completed around the Sun, Saturn made two orbits, and this resonance stabilised the inward migration of both as the gaps they have both opened up in the protoplanetary disk briefly overlapped.

Computer simulations tell us that at this point the interaction of the two planets and the material left in the disk acts to set the process of migration in reverse. Jupiter and Saturn moved outwards, quickly returning to the outer solar system, and moving beyond the region where they had formed. The inner Solar System, particularly the region which today contains Mars and the asteroid belt was left to recover.

It's possible this is why we don't have a mini-Neptune or a super-Earth; so much matter was scattered out of the system by the brief inward march of the giant planets that no such world could form. It is Saturn that made all the difference, and we can give it credit for saving the inner Solar System by stopping Jupiter moving any closer to the Sun.

Throughout this process, Jupiter and Saturn would each have acquired their own circumplanetary disks, within which their remarkable families of moons would later form. Saturn's famous rings came later; most likely they only appeared between ten and a hundred million years ago, long after this drama in the outer Solar System was over.

Further out in the disk, the slightly smaller Uranus and Neptune will have been forming too. They most likely began life closer to the Sun than they are now, and managed to collect up to a quarter of their mass in the gas from the disk. Their remaining mass is made up of rock and ice from the outer regions of the protoplanetary disk, where it was cold enough for molecules including carbon dioxide, ammonia and methane to freeze. This composition explains why they are known as the ice giants – they're different from the gas giants Jupiter and Saturn.

We still know relatively little about these two worlds, despite their proximity compared to exoplanets. Voyager 2 completed a brief flyby of each, but they have not been visited since.

What we do know tells us that these worlds, too, had an interesting early life. The pole of Uranus points toward the Sun, and it rolls around its orbit like a barrel. This may have been caused by interaction between the forming planets, and it's even believed that Uranus and Neptune may have swapped places as Jupiter and Saturn pushed them outwards. Whatever happened, it took time for the four giants to settle down into their familiar positions.

It seems clear that the outward migration of Neptune had consequences for the small bodies in the Kuiper belt, our Solar System's outer asteroid belt. The Kuiper Belt twenty times wider than the more famous asteroid belt that exists between Mars and Jupiter, and it hosts most of our dwarf planets, including the Pluto-Charon system. But the Kuiper belt is different; it seems to lack smaller members, which may have been scattered by Neptune's migration. If so, then this might be the event responsible for the Late Heavy Bombardment, the period – hinted at by studies of the age of craters on the Moon – when the inner Solar System was suddenly and for a period of a million years or so subjected to a huge influx of debris.

Speaking of the inner Solar System, by this point the Earth and the other terrestrial worlds will be nearly fully formed. As with the giant planets in the outer Solar System, their story begins with colliding dust grains slowly building up to larger and larger clumps. The scattering of material in the inner disk by the giant planets, and the increasing instability of the Sun soon removes the raw material, though, halting the formation process. The immediate result is perhaps a dozen bodies the size of Mars or slightly larger, all sharing the crowded space near to the Sun.

In such an environment, collisions must have been common, and something the size of Mars is indeed thought to have hit the newborn Earth between 50 and 150 million years after planet formation started. The impactor is usually called Theia, after the mother of the Moon, because it is in the aftermath of this collision that our planet's companion forms. Studying the composition of the Earth and Moon suggests the collision must have been nearly head on; the material from both bodies was well mixed before it condensed to form the planet and satellite that we know today. One theory is that only a head-on collision could form a large moon like the Earth's; if this is true, we might expect moons like ours to be rare.

Venus too must have been bombarded. A collision with a large body may have slowed its rotation, making its day longer than its year, and collisions between smaller bodies may also explain features on Mars and Mercury. Mars has a significant difference in terrain between northern and southern hemispheres, almost as if significant resurfacing had followed an impact. And Mercury is denser that you'd expect, which might be explained by making today's Mercury the core of what was once a much larger body, its outer layers

▶ **The Solar System**

Artist's impression of our Solar System looking towards the Sun from in the asteroid belt out beyond Neptune.

▶ **The Solar System**

Artist's impression of our Solar System looking towards the Sun from inside the asteroid belt between Mars and Jupiter, and out toward the distant orbit of Neptune and beyond.

lost through an impact. The present state of the Solar System is not the result of neatly ordered processes, as we might once have thought, but rather of the drama of a planet formation process that is more chaotic than anyone imagined.

Our Solar System today

The description of the early life of our Solar System we give above is based on rapidly changing and occasionally controversial science. With each discovery of a world around another star, we learn more about our home. Yet we can be sure that our Solar System is a remarkable place. Its eight planets, along with the moons, dwarf planets, asteroids, comets and interplanetary material, represent a diversity of possibilities that is, as yet, unmatched in our exploration of the Universe. Let's take a tour, starting close to the Sun.

Mercury orbits the Sun in just 88 days, though that is still markedly longer than most of the exoplanets we have discovered to date. It is a small airless body – smaller than several moons of the outer planets – composed of rock with a large iron core that makes up over half of its volume. Even though it exists close to the scorching Sun, it has the wildest temperature swings of any world, due to its lack of any atmosphere. Under direct sunlight, the temperature reaches 700 K, but in the shadows, it could easily be 100 K (-173 C). Some of those shadows are permanent, so they may be cold traps on this hottest of planets – places where ice can exist.

Next up is Venus, the second planet from the Sun. Venus is similar in size to the Earth, but has a much thicker atmosphere comprised primarily of carbon dioxide, which contributes to the enormous pressure at its surface. This thick atmosphere also means that this is the hottest planet in the Solar System, and any explorer unwise enough to land on Venus would be melted, squashed and – thanks to the acid in the atmosphere – dissolved in short order. The presence of hotspots identified in infrared imaging, and the relative lack of craters on its surface, suggest that this is a volcanically active world, though it appears to lack the plate tectonics which continually resurface the Earth.

Earth, you know about. Its main distinction – apart from the large Moon that really makes it more of a double planet – is that it is the only place in the Universe that seems to have life, and possibly even intelligent life. More on that in the next chapter.

Mars, the final rocky planet of the Solar System, rounds out the inner worlds. These days, it has several active robots on its surface and in orbit around it, studying its surface for signs of past life. The red planet is particularly interesting in this respect because we now know it was once a wet world, and liquid water may still exist in briny pockets under the surface. Its once thick atmosphere is now thin, so any astronaut on the surface would be unaffected by even the fastest winds. The winds do pick up and distribute Martian dust, at times creating global dust storms which temporarily change the appearance of the whole planet.

Moving outwards from the Sun, we next meet the asteroid belt which divides the inner and outer solar system. Its nearly uncountable occupants are pieces of rock that range from the size of a car to things the size of a house, with a few more massive inhabitants thrown in for good measure. The gravitational pull of Jupiter shapes the belt, and keeps things in order, though other families of small bodies, including so-called near-Earth asteroids, do exist elsewhere in the inner Solar System.

Jupiter is a true gas giant, making up very nearly three-quarters of the Solar System's planetary mass. It is a heavyweight, being over three hundred times the mass of the Earth

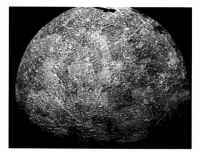

▲ Mercury's southern hemisphere
Mariner 10 was the first spacecraft to fly by Mercury, in 1974. From 440 miles above the surface, hundreds of craters are visible.

and eleven times its size. Composed mainly of an atmosphere of hydrogen and helium, it has nothing we'd recognise as a solid surface; what we see are the tops of staggeringly vast cloud structures formed from ammonia ice crystals. The images returned of Jupiter's clouds by NASA's Juno probe, which has repeatedly swooped close over the giant planet, are worthy of display in any art gallery. Jupiter's retinue of moons also demands attention, from volcanic Io to ice covered Europa and Ganymede.

Saturn is the most distinctive of the Solar System's treasures, thanks to its incredible rings. While each of the giant planets has some sort of ring system, Saturn's is by far the most spectacular. The rings host many moons and moonlets, which shape the bands and gaps that can be seen in images, just as planets sculpt material in protoplanetary disks. Saturn, too, has many moons, the most important of which is probably Titan, a moon the size of a planet with a thick methane atmosphere. Titan has been visited by the Huygens probe, dropped off by the Cassini orbiter which then spent a decade exploring the Saturn system. Huygens revealed a world of rivers and lakes in which hydrocarbons play the role that water does on Earth. Cassini also discovered fountains of salty water spraying from the south pole of Enceledaus, a small icy moon not much larger than the British Isles which must harbour an internal ocean, and which may be a potential home for life.

Uranus is the first of the Solar System's two ice giants, around four times the size of the Earth and thirteen times the mass. Like Jupiter and Saturn, it has an atmosphere which is mostly hydrogen and helium, but it also has a significant amount of methane,

▲ **Venus**

Seen from the Japanese Space Agency (JAXA)'s Akatsuki orbiter, this image uses two ultraviolet wavebands of the atmosphere to discover the composition of Venus's clouds. Three new missions from NASA and ESA will explore Venus in the next decade.

◀ **Mars**

Seen through the eye of the Hubble Space Telescope on March 3 2012 when Mars was in as close to the Earth as it can get, in 'opposition', the ice caps are prominent. The dark feature at the right hand limb of the planet is called Syrtis Major Planitia, and was used by Christiaan Huygens to measure the rotation rate of Mars in the seventeenth century.

▲ Jupiter

Jupiter imaged from the Juno spacecraft. Above the cloud tops is a prominent dark horizontal belt containing a white oval cloud, and a white zone cloud, both of which circle the planet. The Great Red Spot looms on the upper right.

▶▶ Turbulent clouds in Jupiter's north temperate belt.

marking it out as a denizen of the outer solar system, along with Neptune, the furthest planet from the Sun. Neptune is a deeper blue than the blue-green of Uranus, the result of the interaction of methane with the little sunlight that reaches this far. Neptune is also a record breaker, hosting the fastest winds in the solar system at 1,800 mph (2,900 kph). Triton, its largest moon, is likely a captured Kuiper belt object, a strange world with a surface reshaped by ongoing ice volcanism.

We have already mentioned the Kuiper belt which marks the edge of the main Solar System. Pluto and Charon are its most famous members, and while Charon is smaller than Pluto and often considered a moon, it is really large enough that this should be considered a binary system. The centre of mass of the system lies between the two, rather than within Pluto, for example. Pluto's surface, with mountains sculpted by water ice and

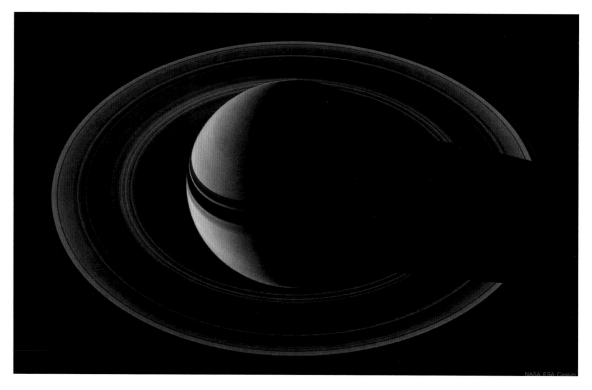

NASA ESA Cassini

▲ **Saturn**

This image of the crescent Saturn in natural colours was taken by the Cassini spacecraft in 2007. The shadows of the rings on the planet and the shadow of the planet on the rings are dramatic features. If you look carefully you can see the moons Mimas (at 2 o'clock) and Janus (at 4 o'clock)

an adorable heart shaped feature, was revealed in exquisite detail by the New Horizons mission which flew past before heading out into interplanetary space. Pluto is joined by the dwarf planets Eris, Haumea, Makemake and many more. The smaller bodies of the Kuiper Belt include Arrokoth, a tiny snowman-shaped object, which was also visited by New Horizons on its way out of the Solar System.

Beyond the Kuiper belt, but still under the influence of the Sun's gravity we have the Oort Cloud, which is thought to span from 300 to 10,000 AU, and which contains at least a trillion comet-like objects. Interaction between objects in the Oort cloud, or perhaps disruption from passing stars, will periodically send some of these objects into the inner solar system; the Oort Cloud is the reservoir which supplies us with long period comets.

The most famous comet is, of course, Halley's, which will be back once more in 2061. At the moment it is too faint to be seen in anything but the largest telescopes, but it will brighten ahead of its next passage through the inner Solar System. Our Solar System also hosts the occasional visitor; along with interstellar dust grains we have detected two larger bodies from amongst the stars. The first, Oumuamua, was bizarre, a small tumbling rock, but the second, Borisov, behaved like a normal comet. More will undoubtedly be found soon.

After that brief tour of the Solar System, it is time to zoom in on the Earth and look at the evolution of our world.

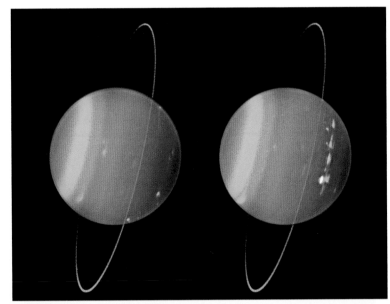

◀ **Uranus**

These images of Uranus at infrared wavelengths were obtained by the Keck telescope in 2004. The north pole is at 4 o'clock. In both, high, white cloud features are seen mostly in the northern (right hand) hemisphere, with medium level cloud bands in green and lower level clouds in blue.

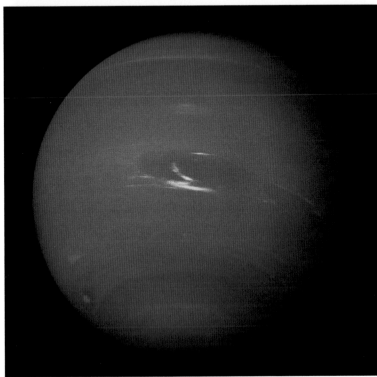

◀ **Neptune**

This picture of Neptune was produced from the last whole planet images taken by Voyager 2 at a range of 4.4 million miles from the planet, 4 days and 20 hours before closest approach on August 25 1989. The picture shows the Great Dark Spot and its companion bright smudge; on the west limb the fast moving bright feature called Scooter and the little dark spot are visible. These clouds were seen to persist for as long as Voyager's cameras could resolve them. North of these, a bright cloud band similar to the south polar streak may be seen. More recent observations suggest the Great Dark Spot has faded.

▼ Triton

The Voyager 2 spacecraft took this colour photo of Neptune's moon Triton on August 24 1989, at a range of 330,000 miles. It is thought to be a captured Kuiper belt object. Regions that are highly reflective in the ultraviolet appear blue in colour. In reality, there is no part of Triton that would appear blue to the eye. The bright southern hemisphere of Triton, which fills most of this frame, is generally pink in tone, as is the even brighter equatorial band. The darker regions north of the equator also tend to be pink or reddish.

▼ **Pluto and Charon**

The New Horizons probe took this image of Pluto
(right) and its moon Charon using data captured
over three days of its close approach in July 2015.

Comet McNaught
Seen from the European Southern Observatory's Paranal Observatory in Chile. The fan-like appearance of the comet's tail is reminiscent of an aurora.

Quadrantid meteor shower

As Earth orbits the Sun, it passes through debris left by visiting comets at precisely the same time each year. This appears as streaking meteors as the debris burns up in the atmosphere. Seen here is the first major annual meteor shower of 2020, the Quadrantids, so called because they appear to originate from the constellation Quadrans Muralis, and enter the atmosphere almost in vertical direction to Earth's orbit

9.2 BILLION YEARS A.B. (AFTER THE BANG) TO THE PRESENT DAY (13.8 BILLION YEARS A.B.)

CHAPTER 5 **The Emergence of Life**

▲ Stromatolites

The nearest thing on Earth to living fossils, stromatolites are rocks made from millions of microscopic layers comprised of the remains of bacteria. They can still be seen in Shark Bay, Australia.

▶▶ Earth from orbit

Astronaut Andrew Morgan took this wide-angle photograph from the International Space Station (ISS) in August 2019. The Soyuz capsule is at bottom, and the ISS's Canadarm2 is on the top. The 16 millimetre fisheye lens, though infrequently used, allows for a unique view encompassing landscapes from two continents—the Nile Delta in Africa and the Sinai Peninsula and the Levant in southwest Asia. The Nile Delta, the grey/blue area near Soyuz, formed where the lower Nile flows north into the Mediterranean Sea. This fertile, vegetated region makes a sharp contrast with the surrounding desert; it has been the centre of agriculture in the region for thousands of years. East of the delta, the Sinai Peninsula acts as a land bridge between the African and Asian continents. South of the Sinai, the Red Sea separates the Arabian Peninsula from Egypt.

▶ Volcano

Mount Erebus, the most active volcano in Antarctica, as photographed by Patrick Moore. Earth's early oxygen-free atmosphere is believed to have formed as a result of a period of prolific volcanic activity. Antarctica was once much closer to the equator than it is today.

A t around 4.6 billion years before the present, the Earth had finally formed, but it was still completely molten. As we saw in the previous chapter, before the surface could cool, a dramatic event took place that resulted in the formation of the Moon. We have already seen that the early years of the Solar System's existence were a lot more chaotic than had previously been thought. The scattering of objects in the Kuiper belt into the inner Solar System, caused by the migration of the giant planets during their formation, resulted in a long period of increased collisions for the rocky planets that were still forming. Eventually, things became quieter, leaving Earth and the Moon as twin worlds. The stage was set for life to appear.

Our planet: cradle for life

During the first 500 million years of its existence, the Earth was too hot to harbour life. The planet's original atmosphere was made up largely of hydrogen captured from the solar nebula, but this could not last. At the temperature of the early Earth, the light hydrogen atoms will be moving rapidly. So rapidly, in fact, that they will quickly exceed the planet's escape velocity, and they will break away from the Earth's grip, escaping into space. Helium atoms are more massive than hydrogen atoms, so they move more slowly, but still fast enough for most of the helium to escape. The little helium that remains in the atmosphere today is just a small fraction of the original primordial atmosphere.

As this was happening, the Earth cooled, forming a solid crust for the first time. The atmosphere was lost so quickly that perhaps there was a brief period where this newly solid Earth had little atmosphere at all. However, the early Earth was a volcanically active world, with eruptions both more common and more violent than they are today. This volcanic activity soon released substantial quantities of gases previously locked up in the rocky mantle and crust, replenishing the atmosphere.

The bulk of the atmosphere you are breathing now is composed of nitrogen which may have been released by this early process. Nitrogen, like many atoms, comes in many

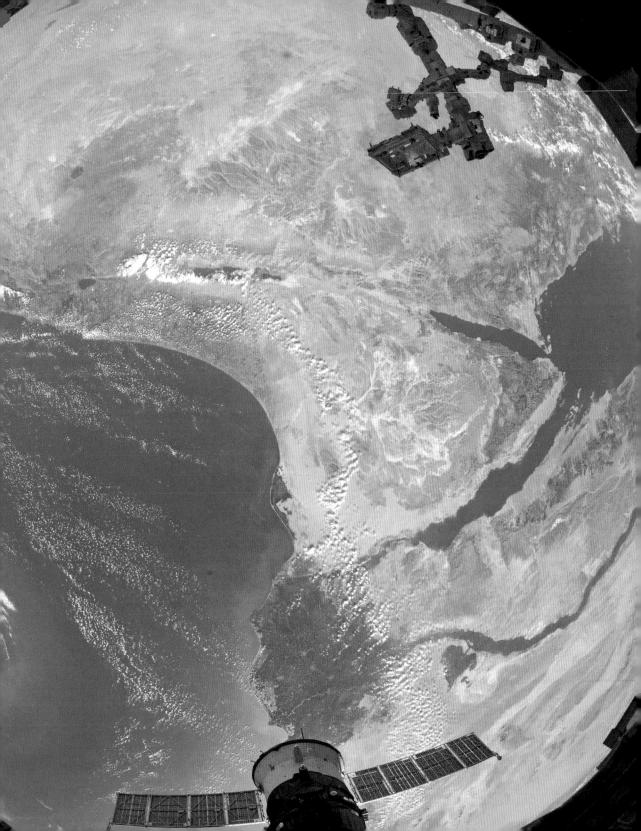

different forms, called isotopes. Though every nitrogen atom must have seven protons, each isotope has a different numbers of neutrons. Chromium nitrate molecules found in meteorite samples here on Earth contain nitrogen with the same proportions of isotopes as found in the Earth's atmosphere. As meteorites like these are the leftover building blocks from the early Solar System, this discovery suggests that the nitrogen in our atmosphere today came to the Earth trapped in the rocks from which our planet formed.

Like our atmospheric nitrogen, it seems that Earth's water came to the planet early in its history. Its source is still a matter of some debate. Some asteroids have more than ten per cent of their mass in the form of water, and it is possible that their impact could have filled the Earth's water supply. Another possibility is a large influx of comets, which are, after all, just dirty snowballs, but the recent Rosetta mission which visited Comet Churyumov–Gerasimenko found that the amounts of deuterium relative to hydrogen did not match those in the oceans. Comets like this one, at least, cannot have contributed much water to the early Earth. More exotic theories exist, including one possibility that water is the result of early life: purple bacteria, found in the layers of stratified lake beds which lack oxygen, produce water through a form of photosynthesis that relies on sulphur instead of the more normal oxygen. If these purple bacteria's distant relatives appeared on the early Earth, they may have contributed the water that made the planet more suitable for the life that we see around us today.

Wherever the water came from, as the atmosphere cooled, the first clouds formed. Then, in a period that we might call the Great Rains, the first raindrops fell to Earth and the lowest lying regions became the first oceans. Water was part of the Earth's story from the beginning, though there may have been significant enhancement during the violent period in the Solar System's history when Uranus and Neptune were migrating outwards, about four billion years ago. Their scattering of Kuiper Belt objects results in the worlds of the inner Solar System being bombarded by smaller, icy bodies. There is also evidence for a 'Late Heavy Bombardment' which seems to have taken place 600 million years after the Solar System formed. This evidence is written on the lunar surface, where most of the craters seem to have been produced during this period. New analysis of the samples brought back from the Moon by the Apollo astronauts, carried out using 21st century laboratory equipment, is shedding new light on this part of our history. We have had to go to the Moon for good evidence, because though the Earth was equally a target, thanks to weather and the movement of the tectonic plates the scars are long gone.

With the raw materials in place, the stage was set for life. Life emerged much earlier than was once believed. The first organism able to reproduce itself probably appeared around 4.3 billion years ago; the isotopes of carbon found in minerals called zircons which are nearly that old show evidence of biological activity. The earliest fossils themselves

Panspermia

The late Fred Hoyle, and his colleague Chandra Wickramasinghe, building on conjecture by the Swedish scientist Arrenhuis, maintained that comets could dump viruses into the upper atmosphere, causing widespread epidemics. (Viruses are strands of DNA or RNA that use the apparatus of living cells to replicate. Some biologists dispute that they are actually alive in the classic sense.) Again there has been little support for this idea, and the idea has never been taken seriously by medical experts.

Probably the most unusual serious theory in connection with life from space was proposed by no less an eminence than Francis Crick, co-discoverer of the double-helix structure of DNA. With the come from between 3.8 and 3.5 billion years ago, a time known as the Archean epoch; we can be reasonably confident of the age, because geologists can work out the age of the rocks in which these primitive organisms are found.

The exact process by which life arose is still unclear; contrary to popular myth, no one has yet come close to repeating this feat in the laboratory. The unproven theory is that chemical reactions were driven by energy from sources such as lightning strikes and short-wave radiation from the Sun. As time went by, more and more complex molecules were produced, until eventually a molecule appeared which could replicate itself. The ability to replicate, or reproduce, is fundamental to anything that we think of as life. The replication was not perfect; each generation brought with it a chance of random variations – errors in the copying process. Some of these random mutations, as they are called, were more successful, surviving longer or reproducing more easily than others, and so these were more likely to form the next generation. This competition between slightly different forms – natural selection – is at the heart of what has become known as evolution. The long stately process which must have led from these simple replicators, no more than complex molecules, to the vast variety of life that we see around us today, had begun.

Under pressure

We have found that life is extremely versatile; some forms are amazingly tolerant, and can thrive in the most unlikely places. The Earth's habitable biosphere extends down ten miles (17 km) below sea level, and up to as high as 40 miles (70 km) in the atmosphere (the edge of space is often considered to be 60 miles (100 km) up). It is thought that one of the first places life may have appeared is around what are termed hydrothermal vents, often known as black smokers. These are fissures in the floors of the oceans that leak out hot, acid-rich water from below; the effluvia are often black – hence the nickname. The temperature of the water emerging from these fissures, at least a mile under the sea's surface, may be as high as 400 °C. Water is able to reach this temperature, higher than its normal boiling point, due to the pressure, which is 25 times that due to the atmosphere at sea level. Remarkably, the fissures are teeming with specialized life forms such as tubeworms, shrimps and even clams, which survive in an environment as acidic as vinegar that would be instantly fatal to most other forms of sea life, and without receiving any energy from the Sun. If life exists under the icy shells of the moons of Jupiter and Saturn, it may well have begun in a similar way; perhaps, in that case, the Sun's light isn't as essential to life as has been thought.

Back on Earth, even in less exotic locations than hydrothermal vents, life in the Archean times would have been very different. Most organisms probably lived in the hot oceans which covered the Earth, and for the first billion years of life's history there was no oxygen

chemist Leslie Orgel, he put forward the theory of 'directed panspermia', according to which life was deliberately sent to Earth by beings from an advanced technological civilization from far across the galaxy. It was pointed out that the chances of micro-organisms being passively transported from world to world across interstellar space were slight, but things would be different if preparations were made. Different types of micro-organism could be carried in a spaceship and deposited here to flourish and develop. When the theory appeared, it is perhaps fair to say that scientists in general were stunned rather than enthusiastic, but ideas of this kind are extremely difficult to disprove.

Comet 67P/Churyumov–Gerasimenko
Mosaic of four images taken by the Rosetta
spacecraft's navigation camera (NAVCAM) on
19 September 2014 at a distance of 28.6 km
(17.8 miles) from the centre of comet 67P/
Churyumov–Gerasimenko. On 12 November
2014 the Philae lander touched down on
the surface of the comet, the first ever such
landing by a probe from Earth.

▲ Black smoker

More correctly known as hydrothermal vents, this
one was photographed at a depth of 9,685 feet
(2,980 metres) on the Mid-Atlantic Ridge. Despite
the darkness, high temperatures and extreme acidity
caused by the superheated gases that are escaping
through these vents, a huge variety of organisms
survive in this hostile environment.

at all. Instead, our atmosphere was rich with methane and carbon dioxide generated by
ongoing volcanic activity. However, at some point this all changed, in what is grandly
known as the Great Oxygenation Event. The instigators are thought to be tiny organisms
known as cyanobacteria – blue-green algae – which evolved to photosynthesize,
turning light, water and carbon dioxide into sugars and oxygen. The carbon dioxide rich
atmosphere provided the conditions these bacteria needed to reproduce rapidly and to
spread throughout the oceans, making the most of the advantage over other bacteria that
evolution had handed them. Rock-like formations known as stromatolites, built by the
bacteria, dating back 3.5 billion years still survive, bearing witness to these tiny creatures'
role in transforming our planet. Living colonies of cyanobacteria can still be seen today,
most notably in parts of Australia's Northern Territory, the living relatives of the ancient
species which gave us a breathable atmosphere.

Worldwide, the fossil record allows us to trace the evolution of living creatures.

Speaking very generally, life evolved rather slowly; for a long time it was confined to the sea, and only during what is termed the Ordovician period, around 450 million years ago, did life spread to the land – land plants first, then arthropods (such as insects, spiders and crustaceans) and vertebrates. It is thought that this migration to the land might be linked to the role of the Moon and the pull of the tides on Earth.

▲ Colony of tubeworms

Relatives of these creatures live near black smokers in extremely hostile conditions. They have no mouth or stomach and survive by absorbing chemicals in the water through their skin. The fish is a vent fish.

The role of the Moon

Our Moon appears to be rather special, and it plays a number of essential roles in the evolution of life on our planet. Its presence and gravitational pull stabilises the tilt of the Earth's axis, which stays within a degree or so of its present value of around 23 degrees. If the Moon was not present, this tilt would vary markedly, leading to shifts in the climate. Compare Mars, which has only two tiny moons, Phobos and Deimos, both captured asteroids: they are too small to provide a stabilising force, and the tilt of the Martian axis varies from 11 to 35 degrees over a cycle of roughly 100,000 years. The Earth's tilt does vary on a shorter 41,000 year cycle but, critically, not by much. It is easy to imagine how the evolution of life – and of complex life in particular – might depend on the long-term stability of the climate, and on the Moon for the stability it grants us.

The Moon's most obvious effect on the Earth is that of the tides that it raises. The tides themselves have important consequences, because they cause friction which slows the Earth's rotation. Earth has a similar effect on the Moon, but because Earth's mass is eighty times larger than that of our satellite, this tidal influence has been much more extensive. Long ago, the friction slowed the Moon's rotation until it became 'captured', or synchronous, so that its period of spin equals its orbital period. As a result, the Moon always presents the same face to the Earth. It is important to remember that this does not mean that the Moon keeps the same face to the Sun; the idea that there is a dark side of the Moon is completely wrong. Day and night conditions on the Moon are the same on the near and far sides, though if you were to live on the far side you would never see the Earth.

The speed of the Moon's rotation quickly became a constant, but the speed with which it travels along its elliptical path never has. Following the usual traffic laws set out by gravity, the Moon moves fastest when near perigee, its closest point to the Earth, and slower when further away. Therefore the position in its orbit and its rotation do become slightly out of sync, and the result is that, as seen from the Earth, the Moon seems to rock to and fro. Sometimes we see a little further around the western limb, and sometimes a little around the eastern edge. Altogether, due to this and other, smaller, librations, as the wobbles are called, we can from the Earth examine a total of 59 per cent of the Moon's surface; only 41% is hidden from us permanently, only to be observed by spacecraft and astronauts who enter lunar orbit.

After the collision between Theia and the early Earth, the newly-formed Moon would have been on a much tighter orbit than it is today, and the tides would have been correspondingly greater. These larger tides may perhaps have created the conditions where life could move onto land; it has been argued that the timing of this great step may have been linked to the changing tidal ranges produced as the Moon receded, under

▶ **Our Moon**

Full Moon photographed by Jamie Cooper on 6 May 2020 using a digital single lens reflex (DSLR) camera attached to an 8-inch Newtonian telescope.

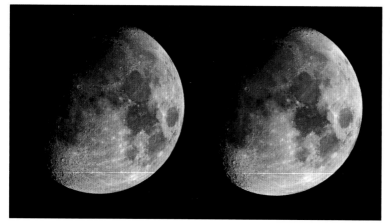

◀ **Our Moon in 3-D**

The same Moon, two years apart, captured on 19 January 2016 and 25 February 2018, and thanks to libration there is sufficient baseline to give us a monumental stereo image of our own natural satellite. Visit LondonStereo.com for advice on how to view in stereo.

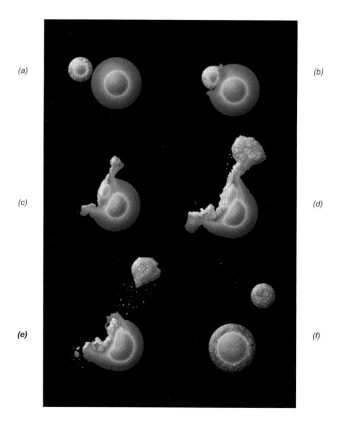

(a)

(b)

(c)

(d)

(e)

(f)

◀ **Formation of the Moon**
A simulation showing the collision of Theia with the early Earth that later coalesced into the Earth and the Moon from the mantle debris.

the influence of the same friction that slowed its spin. The average distance between the Moon and Earth is still increasing at a rate of about four centimetres a year.

Graveyard of the dinosaurs

As life began to take hold on the land, plant growth continued to produce changes to the composition of the atmosphere. Like the early cyanobacteria, plants survive using photosynthesis, which removes carbon dioxide from the air and uses it to build food in the form of sugar molecules. A waste product of this process is oxygen, which plants release into the air, and it is this oxygen which allowed complex animal life to evolve.

Life's progress was not smooth. The greatest catastrophe in life's history occurred at the end of the era known to geologists as the Permian, over 250 million years ago. The Permian lasted for around sixty million years, and seems to have been a time of widespread deserts. Most of the world's land masses were joined together in a vast continent, which has been called Pangaea. It seems that this Permian extinction, often known more poetically as the 'Great Dying', was the most extensive in history and wiped out most of life on Earth. This, of course, can be established by the fossil record, but there is no crater left to guide us as to the cause of the disaster. Instead, we must depend on certain carbon molecules known as fullerenes. These molecules form a cage-like

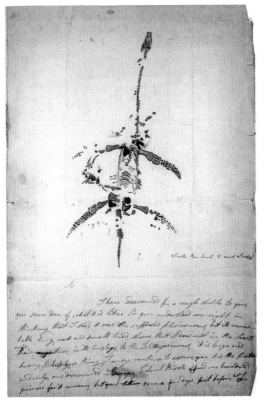

◀ **Plesiosaurus dolichodeirus**

Letter and drawing from Mary Anning (1799–1847) announcing the discovery of a fossil animal now known as *Plesiosaurus dolichodeirus*, 26 December 1823. The letter begins: "I have endeavoured in a rough sketch to give you some idea of what it is like. Sir, you understand me right in thinking that I said it was the supposed *Plesiosaurus*". The fossil dates from around 200~180 million years ago in the Early Jurassic period.

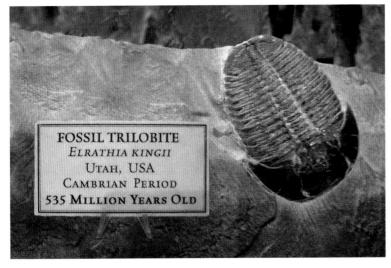

FOSSIL TRILOBITE
ELRATHIA KINGII
UTAH, USA
CAMBRIAN PERIOD
535 MILLION YEARS OLD

◀ **Inspirational history**

It was a book by Patrick Moore, *The Earth*, in the school library, that introduced Brian to the amazing story of the trilobites, and inspired him to a life-long passion for astronomy. There were once 15,000 species of trilobite, and by the time they were wiped out by the Permian extinction 250 million years ago, they had roamed the Earth for 300 million years. By comparison, we humans have so far been around for less than 200,000 years. The trilobites' nearest living relative is the horse-shoe crab. This particular trilobite recently found its way to a nature shop in New York, for sale along with a fine selection of meteorites, dinosaur bones, and other clues to our distant past.

structure, most often in the shape of a ball, and inside this cage single unreactive atoms are trapped at the time of formation. The helium and argon found in fullerenes at the end of the Permian seem to have come from space, produced in the atmosphere of a star that exploded as a supernova before the Sun was formed. These chemicals may be the remains of a meteorite that carried material left over from the beginning of the Solar System. It is suggested that, as a consequence of the impact, there may also have been huge amounts of volcanic activity, covering the entire land surface in lava to a depth of nine feet (3 metres). It is therefore far from surprising that 90 per cent of all marine species and 70 per cent of land vertebrates failed to survive.

Reptiles began to appear throughout the Permian, and we come to the age of the dinosaurs, some of which were huge and ferocious hunters while others were small and herbivorous (plant eaters). One small harmless dinosaur, no larger than a canary, has been nicknamed the Tweetieosaurus.

Dinosaurs ruled the world for almost 200 million years (in comparison, human beings have been on Earth for less than 200 thousand years), but then, at the end of the geological Cretaceous period, 65 million years ago, the great dinosaurs suddenly vanished. Yet the extinction may not have been total; it now seems certain that some of the smaller species survive to the present day in the form of their feathery descendants, the birds. The departure of the large dinosaurs may have been a good thing from our point of view. It meant that mammals could diversify from small shrew-like animals into a much wider diversity of species, eventually producing the apes that evolved in the Miocene period (25 to 5 million years ago) which are our direct ancestors.

Investigation of the cause of extinctions is a popular pursuit, and opinions vary. So far as the demise of the dinosaurs is concerned, the favoured theory of today is that a large meteorite struck the Earth, throwing up a colossal amount of dust, and causing global devastation. The site of the impact has been identified; Chicxulub off the coast of Mexico, in the Yucatan peninsula, where we can detect the eroded traces of a crater 93 miles (150 km) across and 12 miles (20 km) deep. From the shape of the crater, we can even pinpoint the angle with which the fatal meteorite arrived; it came in from the northeast at an angle of 45–60 degrees.

▼ Chicxulub basin

The Chicxulub crater on the edge of the Yucatan Peninsula in Mexico may be the site of the impact that ended the reign of the dinosaurs.

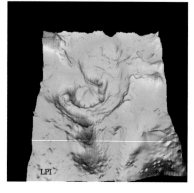

Further evidence is found in rocks laid down at this period over a large part of the Earth, which contain more than the expected amount of iridium – an element comparatively rare on Earth but characteristic of meteorites. We cannot be certain that the impact wiped out the dinosaurs, but the theory is widely supported.

Life in the Universe

Is there life elsewhere in the Universe? In considering this question, we have to make one point clear at once: we are discussing life as we know it. All life we know of is based upon carbon, and it makes sense to assume that life, wherever it exists – here, on Mars, or on some distant planet – is carbon based. Similarly, as discussed above, we assume that life requires water to act as a solvent, so places like the cold, airless Moon can be rejected out of hand. The counter-argument, of course, is that this might be completely wrong, and that there could exist in the Universe, for example, intelligent beings whose bodies are based upon atoms of gold, who breathe in atmospheres of sulphuric acid, or a billion other possibilities for what science fiction used to call

Bug-Eyed Monsters. If they do exist, then we apologise, but making conservative assumptions helps us to begin thinking systematically about life.

Start with temperature, which is critical in determining whether a planet could have liquid water. Around every star there is a region known as the temperate – or 'Goldilocks' – zone. Get closer to the star, and its heat would boil water into steam. Get further away, and water will freeze into ice. At distances in between, where temperatures between 0 and 100 degrees Celsius as possible, water could exist as a liquid. If a planet's orbit around a star keeps it within the temperature zone then – if it has a solid surface – liquid water could pool there, forming lakes, seas or oceans.

We have already seen that most of the exoplanets we have discovered to date have orbits with periods less than 100 days. It might be thought that planets on such orbits, close to their stars, would be far from the temperate zone, but we should consider the types of star involved too. A majority of the stars in our galaxy are small, cool, red stars (those in the bottom right of the HR diagram). These stars might be half the temperature of the Sun, and so you would need to push the Earth much closer-in if it were to receive the same amount of heat as we do in our current orbit. Conversely, if the Sun was a giant blue star (from the top left of the HR diagram) we would surely bake in our present orbit, and we would need to be much further out for our planet to have liquid water. Many of the planets we have found seem to be at the right distance from their stars to lie within the temperate zone.

We can go further, using Earth as our lodestar. Using a model that tells us how Earth's atmosphere has changed through time, we can define not only a temperate zone, but an Earth-based habitability zone within which the Earth would have a climate stable enough for the emergence of life as we know it. This range is called the habitable zone, and in our Solar System it stretches further out than you would think, beyond the orbit of Mars. This is because if we put Earth where Mars is now, liquid water could still exist on its surface. But Mars is, at best, only marginally habitable right now, so how can it be in the 'habitable zone'?

The red planet may once have been a haven for life. There is distinct evidence that there was once flowing water on its surface; images taken by orbiting spacecraft show geological features that match what weathering by water creates on Earth, and the rovers that have explored the planet have found minerals that must have formed in the presence of water. Mars' smaller size and lower gravitational pull means that, unlike the Earth, its atmosphere has been gradually lost. In turn, this means that its ancient seas and oceans would have evaporated. It is also thought that Mars once had a magnetic field that would have shielded the surface from high-energy cosmic rays, but as the interior solidified this field will have disappeared. Mars' small size and its position produced the desert world we observe today; if it had been larger and the Earth smaller our questions may have been reversed.

Having defined the habitable zone using the Earth's properties, we can calculate where its boundaries lie for each star that we study. The Kepler mission, which we met in the previous chapter, was designed precisely to search for worlds in this zone, and it allows us to calculate the occurrence rate for planets – the number of planets per star which have some particular properties. The idea is to get away from worrying about the likelihood with which different types of planet will be detected, and consider the underlying rate at

▼ **The Search for life on Mars continues**

NASA's Perseverance rover is the latest rover to explore the surface of Mars in search of evidence of microbial life having existed in the past by collecting rocks and soil samples. It touched down on February 18, 2021, and amongst its host of ingenious technological innovations is the helicopter Ingenuity. This image of Ingenuity's second flight was captured by Perseverance's MastCam-Z.

▶ **The habitable zone**

The habitable zone is the zone around a star where it is neither too cold nor too hot for liquid water to exist on the surface of a planet. The diagram shows the extent of the zone as it might apply to a planet like our Earth. However, the atmosphere of an exoplanet might be very different to that of Earth, able to retain heat better or to lose heat at a faster rate, so the range of the habitable zone may be very different.

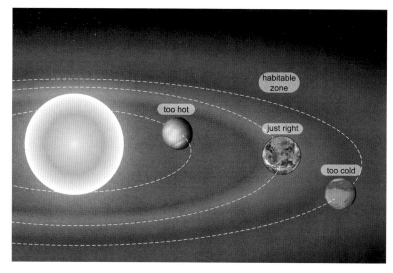

which they exist. Most of the exoplanets found so far in their stars' habitable zones are giants, without any sort of surface we would recognise, but this just may be because they are easier to detect.

The goal is to measure a number called 'eta-Earth': how common are Earth-sized, Earth-mass planets at one astronomical unit away from a Sun-like star. Though we are making progress, estimates in the literature at time of writing still range from 1.3% to 124%, a range which means that somewhere between one in a hundred and pretty much every Sun like star has an Earth of its own.

▼ **Exploring the surface of Mars**

This self-portrait of NASA's Curiosity Mars rover shows Curiosity on Mount Sharp, at the so-called Mjoave 2 site where it gathered samples in January 2015.

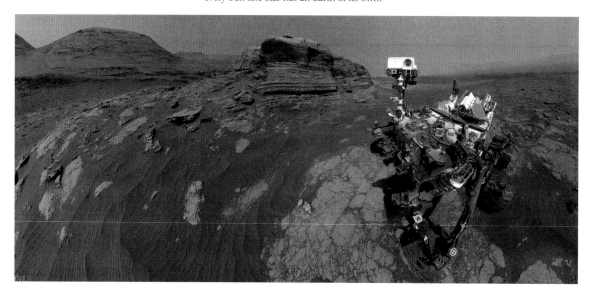

What about other types of star? One of the first studies on exoplanet occurrence rates was conducted not on Sun-like stars but on the small, cool, red stars called M dwarfs, and this suggested the closest known habitable zone world could be no more than 13 light-years away. But worlds such as this would be very different from our own. Life on Earth depends on having a stable star like the Sun, rather than a host which is prone to energetic flares as many M dwarfs are. Life on Earth may also depend on there being sufficient water on its surface, an atmosphere that contains a significant amount of oxygen (itself a product, as we have seen, of life), a stable axial tilt which also produces changeable but not too dramatic seasons, the existence of a planetary magnetic field and perhaps on the presence of tectonic activity. With so many different factors involved, it will be difficult to say which planets may be truly habitable, but by considering the most important factors we can make a start.

Many current attempts to identify truly habitable worlds concentrate on their atmospheres. We know that on Earth, the composition and thickness of the atmosphere was and is a huge factor in making our world home to life. We can view the Earth as we do exoplanets by considering sunlight passing through, or being reflected by, the atmosphere, recording the presence of different molecules which absorb or scatter the light. This is similar to a technique that can be used to study the atmosphere of transiting exoplanets, where starlight shines through the atmosphere as the planet passes in front of its star. If we measure the spectrum of the star before and during a transit, we can hope to find the fingerprints of molecules in the planet's atmosphere. We can also try and find the light reflected from the atmosphere just before the planet disappears behind the star, as seen from the Earth.

Playing this game with the Earth reveals a number of telltale markers that would suggest to an observer that our world is more than just another rock with an atmosphere. The biggest giveaway is probably what is known as the 'red edge', a change in reflectivity that appears as you switch from optical to infrared wavelengths and which indicates that something on the planet's surface has got to be very good at absorbing visible light. The presence of chlorophyll in photosynthesizing plants on the Earth alters how our planet looks. At wavelengths where the plants absorb light – mostly at the blue end of the spectrum – the planet is darkened, but the plants reflect green and red light, making the planet bright in these colours. This red edge can even be seen to change as the Earth rotates, bringing different continents into view. When South America and the still extensive Amazon rainforest is pointed toward an observer, the red edge would appear strong, but it would disappear almost entirely when the Pacific Ocean is front and centre.

If we were looking for a truly Earth-like world, finding a red edge would not on its own constitute definitive proof that life existed. We would also want to look for the signs of absorption by oxygen, the presence of ozone, and features like the imbalance between methane and carbon dioxide which do not appear naturally. We might even be able to detect the damage (nearly) intelligent life can do to a planet; on Earth, the presence of chloroflurohydrocarbons or CFCs in the atmosphere indicates the presence of an industrial civilization that has not yet learned to clean up after itself. Perhaps alien observers are, even now, watching the build-up in carbon dioxide that is changing the Earth's climate with some concern.

The point is that observational signatures of life do, in theory, exist, though they are subtle. What are the odds of a search succeeding? Our Galaxy is filled with around 400

▼ **Christmas on an alien world**

One of the playful visions of what life might be like on another planet by Gertrude Moore, Patrick's mother.

▼ **Saturn from Titan's atmosphere**

NASA's Cassini spacecraft views Saturn through the atmosphere of the planet's largest moon, Titan. The image was captured on 31 March 2005 while Cassini was 7,500 km (4,600 miles) above Titan, and 1.2 million km (0.75 million miles) from Saturn.

billion stars. Let's assume each one has a planet. Keeping only the stars which are stable enough not to throw out lots of harmful radiation in the form of violent flares and coronal mass ejections, we might be left with about 80 billion stars. Assume one in four have a planet in the 'habitable' zone, as defined by the region where a planet with an Earth-like atmosphere might have water. Assuming a suitable fraction of these planets are rocky, current estimates suggest that there are, in the Milky Way, between 200 million and six billion planets where conditions seem to be right for life to form. On how many of these did it actually get started? We don't yet know.

To answer this question, perhaps the most difficult of all, we need to know and to understand the mechanism by which life got started on Earth. If the probability is only one in a thousand billion, making us the winners of some cosmic lottery, then the existence of even our civilization is surprising. If, as some believe, it might be as high as one in a hundred, then we will have many millions of populated planets to search.

One way of constraining this range of possibilities is to look for other places where life may have formed in our Solar System. Candidates range from the upper atmosphere of Venus, to the Martian subsurface where life might cling on, to the oceans on moons of Jupiter and Saturn or the methane lakes of Titan. If we find life of any kind, in any of these places, no matter how simple or lowly, it will favours those who believe that life will appear wherever it can. We will then be entitled to assume that life is common throughout the Galaxy – but will it be worth talking to?

The anthropic principle

In modern cosmology there has been an attempt to examine what are known as 'anthropic arguments'. These are based on the so-called anthropic principle, which states that the Universe must be the way it is, because if it were to be any different, we would not be here to observe it! To give a trivial example, if the Universe were the size of an atom, creatures complex enough to be conscious could not exist. Sophisticated versions of this argument have been developed, and used to support the hypothesis that we really are unique. However, it is difficult to see how such theories may be tested.

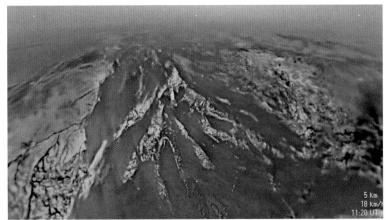

5 Km
18 Km/h
11:20 UT

◀ The descent of Huygens

On December 25, 2004, the Huygens probe was released from the Cassini spacecraft and it arrived at Titan on January 14, 2005. The probe began transmitting data to Cassini four minutes into its descent through Titan's murky atmosphere, then it touched down, the first time a probe had landed on an extraterrestrial world in the outer Solar System.

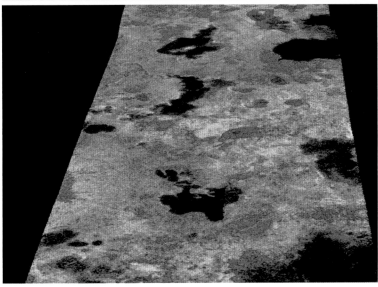

◀ Titan

In the images of Titan's surface from the Cassini probe, some areas reflected very little radar, and the most likely explanation is that the region is dotted with lakes, composed of liquid methane. This map spans about 150 km (93 miles). Apart from Mars, Titan is the only other body in the Solar System to possess liquids on the surface. The Dragonfly spacecraft will soon visit this enigmatic world.

The adventures of Huygens

Titan is the only moon in the Solar System with an atmosphere. The thick orange clouds of methane and other hydrocarbons hid the surface from view until the arrival of the Cassini orbiter in 2004. It carried a lander, Huygens, which in January 2005 plunged through the clouds towards an unknown surface. As the small probe fell, it made measurements of the atmosphere and took photos of a landscape and the conditions. The scene is reminiscent of a dry lake bed, which is exactly what it is – though more recent evidence suggests the site gets occasional rains of methane. Later observations from orbit – once Cassini had learned the trick of tuning its cameras to see through the atmosphere – revealed small, but long lived seas of hydrocarbons, particularly methane and ethane on the surface. Could life exist in such an environment? There is certainly complex chemistry, though it is very different from the chemistry we are familiar with on Earth. Within those hydrocarbon seas, there may be simple organisms which have evolved to make use of methane and the other raw materials; if so, then it will be a sure sign that we should expect a Milky Way filled with diverse life. It may be that while nothing so complicated as a bacterium swims in Titan's seas, there still might be repeated patterns of chemical reactions – almost, but not quite complicated enough to be life. Titan is a glimpse of what the early Earth may have – a prebiotic world rather like the Earth must have been in the Archean.

▲ **The Search for Extraterrestrial Intelligence (SETI)**

In 1896 Nikola Tesla suggested that a version of his early wireless electrical transmission system could be used to contact Mars, and since then many attempts have been made to do so, extending from physical messages onboard the Voyager spacecraft launched in 1971, across the electromagnetic spectrum, using powerful telescopes such as 26 metre (85 foot) Robert C. Byrd Green Bank Radio Telescope which focuses the radio waves falling on it onto sensitive receivers at the top of the boom attached to the side. This was the telescope originally used in 1960 by Frank Drake to run a SETI programme, named Project Ozma after the queen in the film The Wizard of Oz. Frank Drake subsequently became renowned for his "Drake Equation", multiplying together all the factors which were deemed relevant to ascertaining the probability of life existing on a planet elsewhere in the Universe. Despite the fact that we know of more than 5,000 exoplanets, we are no closer to finding life on any of them than we were in 1960, when Project Ozma failed to deliver. And the Drake Equation, as Drake himself admitted, had too many unknowns to be much use. Today, SETI programmes abound, with the prospect of finding life in some form always a motivator for planetary landers withing our Solar System. The SETI Institute in California, established in 1985, and now employing over 100 scientists, provides leadership in SETI research and education.

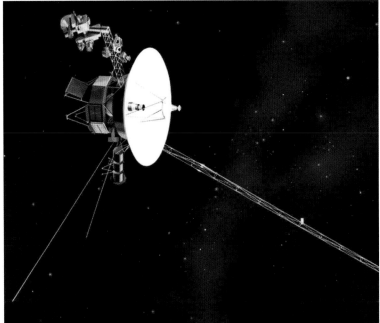

◀ Voyager 2

Launched in 1977, Voyager 2 flew past Uranus (1986) and Neptune (1989), returning the first ever close-up images of these outer planets. It is the first man-made object to leave the Solar System altogether and head for the stars. It carries gold-plated gramophone discs containing images and sounds from Earth, in case it encounters alien life. The records include a schematic diagram showing Earth's position in the Galaxy, so that any recipients of these interstellar gifts will be able to thank us in person.

▼ Exiting the Solar System

In addition to Voyagers 1 and 2, Pioneers 10 and 11 have also left the Solar System. New Horizons is deep into the Kuiper belt heading out of the Solar System.

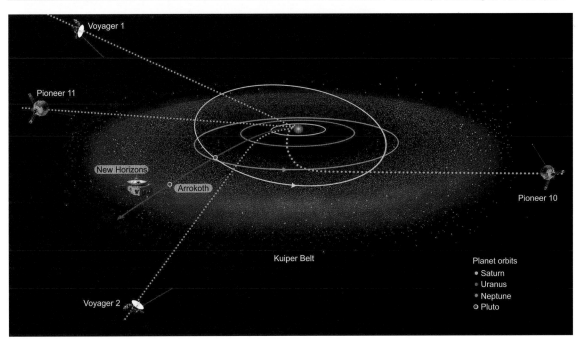

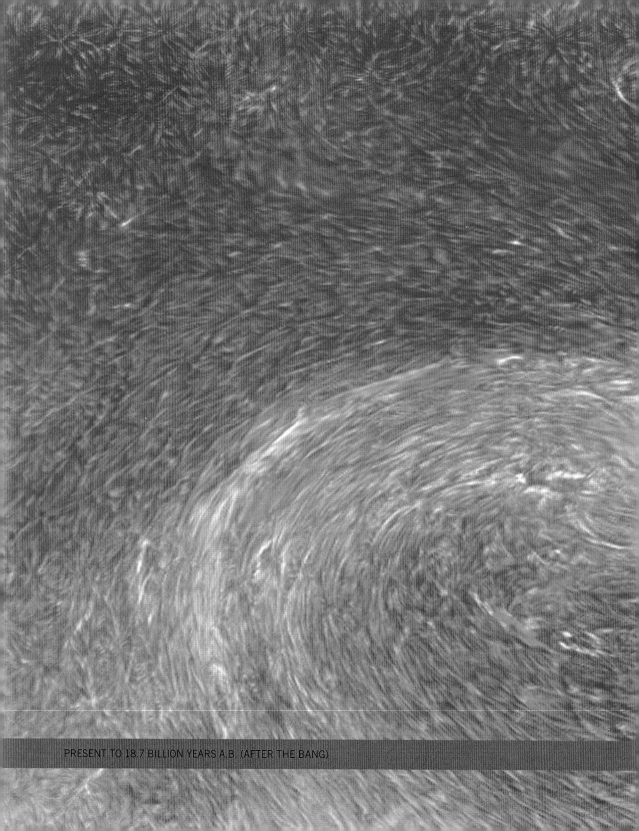

PRESENT TO 18.7 BILLION YEARS A.B. (AFTER THE BANG)

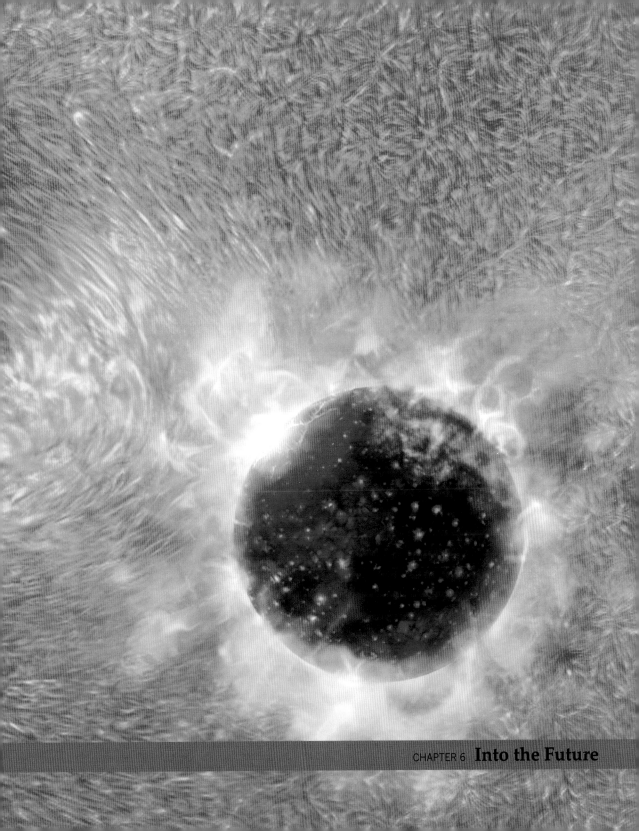

CHAPTER 6 **Into the Future**

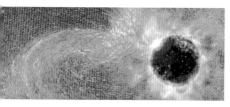

▲ The end is nigh

Five billion years from now, the red-giant Sun will have expanded to such a size that the inner planets Mercury and Venus will be subsumed, and Earth will suffer a fiery death.

When looking into the past we have actual evidence to examine: in the Earth's fossil record we can glimpse the very early stages of our planet's history; in the craters of the Moon we have evidence of ancient cataclysmic meteorite impacts; in the clouds of the Crab Nebula we see the remnant of a violent supernova that occurred almost one thousand years ago. As we gaze at the faint light of the galaxies we are already seeing them as they were millions of years in the past. If we measure the rate at which they are fleeing away from us, we can build up a reliable picture of the state of the Universe as it was billions of years ago; and as we contemplate the cosmic microwave background, we are literally viewing the Universe just 300,000 years after the Big Bang. We can actually see the past.

▶ One of the wonders of nature

Patrick Moore, standing in Meteor Crater near Flagstaff in Arizona. Caused by the impact of a meteorite around 50,000 years ago, Meteor Crater is about 4000 ft(1200 m) in diameter and 570 ft (170 m) deep. Patrick visited for *The Sky at Night* during the 1970s.

▼ Iron meteorite

Brought from China, the meteorite fell in 1516, during the Ming Dynasty.

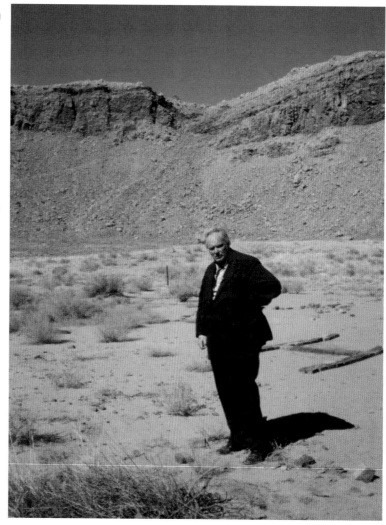

The future is more problematic; we cannot see stars and galaxies as they will be in the future, and we have to rely on deduction, mixed with a good deal of scientific speculation. Though many pages of the history of the Universe have yet to be deciphered, we know much more about the Universe of six billion years ago than we do about the Universe of six billion years hence.

The Earth may be insignificant in the Universe, but to us it is obviously of paramount importance, so let us look first at what may lie in store for our own planet. On average, the Earth is hit every few hundred thousand years by an asteroid large enough to cause widespread devastation. Indeed, recently we have tracked several asteroids that have passed alarmingly near the Earth; a few have brushed by at a distance of only a few tens of thousands of miles, well within the orbit of the Moon.

A collision with a body only a few miles across would be disastrous for humanity, but we are making good progress in tracking asteroids large enough to cause devastation on a global scale. With most large and potentially hazardous objects recorded, and their orbits tracked, we can be sure that none of the known large asteroids will hit the Earth during the twenty-first century. It is hard to predict orbital motions with sufficient precision beyond the next hundred years, and at some point our luck will run out. A range of options – from painting asteroids white to increase the radiation pressure sunlight exerts on them to more dramatic missions – are being developed and tested, and it is not too optimistic to hope that – politics aside – we may be well-equipped to defend ourselves when the next really big impact is due. Of course, there are many smaller objects capable of causing damage to a city or even a whole region; efforts to find, track and catalogue these are underway.

There are other wholly possible natural scenarios in which life on Earth could come to a premature end. Geologists have recently become aware of the potential eruption of

▼ Sumatran supervolcano?

One of the largest known volcanic eruptions occurred 74,000 years ago when the Toba volcano erupted in Sumatra, leaving the approximately 1,000 square mile (3,000 square km) Toba caldera behind. This giant depression formed after the collapse of the volcano's cone, seen in this satellite image (below left) and at ground level (below). The island in the crater lake is a resurgent dome, where magma is active in a chamber below the surface.

▶ **Ice on Mars**

Three images from the Hubble Space Telescope show how the ice cap fluctuates as the seasons change – here the progression (from left to right) is from Martian autumn to spring and then summer.

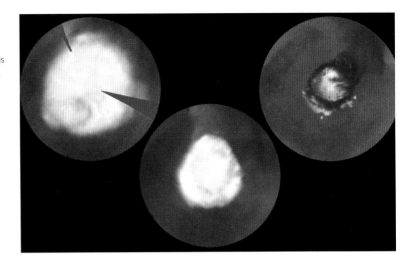

supervolcanoes, which could result from vast reservoirs of magma under extreme pressure, one of which has been discovered under the Yellowstone National Park in Wyoming. The eruption of any of these could result in a planet-wide cloud of debris in the atmosphere, so dense and persistent that most plant and animal life would die from lack of sunlight. It is now thought that some past extinctions may have been due to supervolcanoes.

Humanity is also capable of inflicting immense harm on itself, and on the billions of other species that share our planet. We are changing our planet's climate, and without drastic action the Earth of just a century hence will be a much warmer, and less hospitable place. The planet itself will endure, but these changes have dramatic implications for life as we know it. Such climate change is very fast on an astronomical timescale, and cannot be explained by changes in the Earth's orbit, or by solar activity. Consider the next few hundred million, or even the next billion years, and the reverse is true; our petty squabbles and pollution are irrelevant, and the Earth's long story is dictated by the Sun.

▼ **Arctic sea ice**

The extent of the Earth's north polar sea ice has recently been decreasing at a rate of nine per cent per decade. This picture shows the situation in 2004.

The Sun and the Earth

All through the history of the Earth there have been warm and cold periods. The last Ice Age ended a mere 10,000 years ago; there have been minor variations since then – during the 'Little Ice Age', between around 1645 and 1715, there were times when the River Thames in London regularly froze in winter, and frost fairs were held on it. So what causes these changes in climate?

The Serbian engineer and mathematician Milutin Milankovič attributed them to the movements of the Earth itself. The pattern of our orbit round the Sun is not circular; it is elliptical, and the eccentricity of the ellipse varies between certain limits over a period of just over 400,000 years. At present the Earth's axial tilt (the obliquity) is 23.4° to the orbital plane, which is why we have our seasons, but over a period of around 41,000 years the obliquity ranges between 22.1° and 24.5°; it is decreasing, and will reach its minimum value around the year AD10,000. Precession (the change in direction of the Earth's axis of rotation relative to the stars) also varies, over a period of 26,000 years, and this affects the positions of the celestial poles; when the Pyramids were being built, the north pole star was not Polaris, but Thuban in the constellation of Draco. Taking all these various considerations into

◄ **Tunguska**

In June 1908 something exploded above the Stony
Tunguska river in Siberia. Witnesses reported
seeing a bright fireball. The explosion was heard
600 miles (960 km) away; 380 square miles (970
square km) of forest were flattened; trees up to 30
miles (48 km) away were felled by the shockwave.
It is probable that the explosion was caused by a
meteorite 160 feet (49 metres) in diameter entering
the atmosphere and vaporizing five miles (8 km)
above the ground.

The end of life on Earth

The Sun is using up its nuclear fuel but, surprisingly, becoming more luminous. This
happens very slowly – imperceptibly, as far as we are concerned. As the hydrogen in the
centre of the star is used up, so the Sun contracts a little, putting more pressure on the
core and raising its temperature. The rate at which the reactions proceed depends strongly
on the core temperature, and so fuel is used up faster. A billion years from now, the Sun
will be powerful enough to give Earth an uncomfortably torrid climate; any surviving
inhabitants would surely have abandoned the equatorial regions altogether long before
this point, with the last earthly lifeforms huddling near the poles.

But this will only provide a temporary escape. The deserts will expand as the lower
latitudes become uninhabitable, and the land available for cultivation of crops will become
scarce indeed. The shifting of the continental plates will have long since destroyed the
familiar shapes of the continents. Any remaining ice caps will melt, causing an enormous

account, it was suggested that the 'Milankovič cycle'
could explain the warm and cold spells.

Other investigators disagreed. After all, we depend
entirely upon the radiation we receive from the Sun
and although the Sun is a steady, well-behaved star,
fortunately for us, it is variable to some extent. The
11-year solar cycle is well known, but there are other
factors too. During the 'Little Ice Age' the cycle was
apparently suspended, and there were few, if any,
sunspots. When the spots returned, there was a
period of global warming. Unfortunately our reliable
sunspot records only go back a couple of centuries,
but the link with climate cannot be doubted. When
the Sun is at its least active, greater numbers of
cosmic rays can reach our atmosphere, causing
increased cloudiness and a fall in temperature.
Short term climate change is undoubtedly caused
by our activity.

Eventually, of course, the Sun will run short of its
hydrogen 'fuel' and become a red giant star, with
disastrous effects on its planets, but this crisis will
not be upon us for so long that it need not concern
us. In the foreseeable future we must certainly
expect warm and colder periods, but there is every
reason to believe that the Earth will remain habitable
for at least the next thousand million years.

▶ **Migration to Titan**

While its thick, nitrogen rich atmosphere has put the planet on some people's list of potentially habitable Solar System refuges, if human migrants warmed the atmosphere by a few degrees it may escape Titan's low gravity.

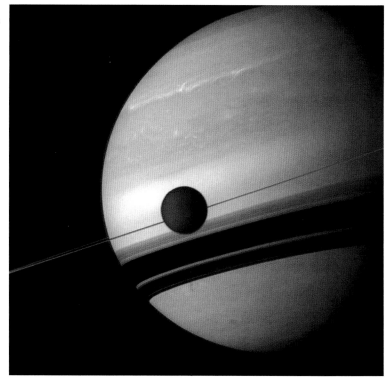

rise in sea level; much of the remaining land will be flooded.

The relentless heat will increase; by three billion years in the future a critical point will have been reached. The Sun will be 40 per cent brighter than it is now, so that all surface water on the Earth will have evaporated; the oceans will be gone, and our world will have become a very hostile place.

If humanity still exists on Earth when the changes in the environment become obvious, how will our remote descendants react? The onset of these changes will be detectable, and alarm bells will ring – but it seems unlikely that even a highly advanced civilization could control the Sun. No doubt a committee meeting would be called, but what would be on the agenda? To move the Earth out to a safe distance might be possible, but even this would not provide a permanent solution, as we shall see. It might be possible to remove the Earth from the Solar System altogether and somehow make it self-sufficient, so that it could survive without a Sun. If this proved too difficult the human race might consider mass migration to another world – in another solar system – or else the construction of an enormous, self-supporting space station to accommodate the survivors.

If nothing can be done, as time passes it seems likely that the entire Earth will become a molten, seething mass of magma. There can then be no reprieve; ultimately all life will be wiped out. So much for Earth. In the rest of the Solar System, things may become temporarily more promising for life. Mars will be much warmer than it is now, and its

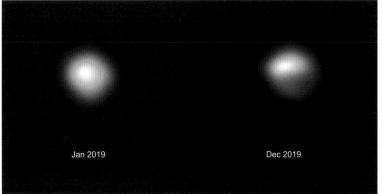

Jan 2019							Dec 2019

◀ **Betelgeux**

Normally one of the ten brightest stars in the night sky, Betelgeux began getting dimmer in 2019, and by mid-February 2020 had lost more than two-thirds of its brilliance – a difference noticeable to the naked eye. The star has now brightened once again.

massive ice caps (composed of both carbon dioxide and water) will begin to melt. An atmosphere will develop, and for a short while – a few tens of millions of years or so – Mars will briefly be a hospitable place. However, this situation cannot last for long. Mars is simply too small, and has too weak a gravitational pull to retain its newly found atmosphere for long.

It has been suggested that humanity might find a refuge on Titan, the largest satellite of Saturn, which has a thick, nitrogen-rich atmosphere. Alas, this is not so. Titan has a low escape velocity, and retains its atmosphere only because it is so cold; at a low temperature, gas molecules are sluggish. Raise the temperature by only a few degrees, and the whole of Titan's atmosphere will escape.

During the following half a billion years the Sun will swell to over twice its present size, and although the surface temperature will fall, its luminosity will double. There will also be effects on the Earth's orbit. The Sun's stellar wind will increase in power and our star will begin to lose mass as it evolves into a red giant. This loss of mass means that the Sun's gravitational pull will be weakened and, in response, planets will start to move outward; the Earth will swing out to a distance of around 120 million miles (200 million km) – not nearly far enough for it to escape from the intense heat of the now massively-swollen Sun.

The red-giant Sun

Moving further into the future, at about five billion years from the present day, hydrogen 'burning' in the Sun's core will cease; there will be no hydrogen left – it will all have been converted to helium in the process of nuclear fusion. The core suddenly will no longer be supported by the pressure of radiation emitted from nuclear reactions. Gravitational collapse cannot be prevented; the outer material will rush in, compressing the core and heating the material. Until this point, helium nuclei had been unable to participate in nuclear reactions. In a matter of seconds, however, the temperature will become high enough for another level of fusion to occur. The helium nuclei combine to form beryllium and lithium. This is a much more efficient reaction; the Sun will radiate over 2000 times as fiercely as it does now and it will balloon out to such an extent that Mercury and Venus will be swallowed up. The Sun will have, eventually, become a red giant.

▼ **Swallowed by the Sun**

The Sun will swell to such a size that Mercury and Venus are consumed. While the current orbit of Earth will be within the red giant, the star's loss of mass will cause Earth to swing outward and so escape. By then, life on Earth will long have ceased to exist. The diagram below shows the size of the swollen red-giant Sun, compared to the size of the inner Solar System today.

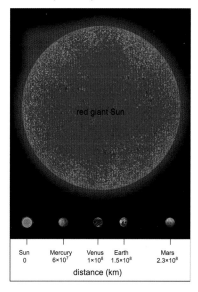

red giant Sun

Sun	Mercury	Venus	Earth	Mars
0	$6×10^7$	$1×10^8$	$1.5×10^8$	$2.3×10^8$

distance (km)

At some stage in its evolution, the ageing red-giant Sun will become increasingly unstable. Its outer envelope will be blown to a distance from the main star by a series of violent pulsations, forming what is known as a planetary nebula.

It is worth noting that a planetary nebula has nothing to do with planets, but is simply the discarded outer envelope of a highly-evolved star. These are the butterflies of the Universe, with many beautiful and varied forms but with lives of only a few tens of thousands of years. The most famous of these objects, the Ring Nebula in Lyra, is easy to locate even in a small telescope, because it is midway between two naked-eye stars, β (beta) and γ (gamma) Lyrae, close to the brilliant Vega; even moderately powerful binoculars will show it. In a telescope it looks rather like a dimly luminous cycle tyre. M57 looks symmetrical, but other planetaries show an amazing range of shapes, which must depend on the exact processes by which material is ejected from the central star;

The future of the Moon

The Moon will remain linked with the Earth – there is no reason to suppose otherwise – but its orbit will change. At the moment it is moving away from the Earth at the rate of about 4 centimetres ($1\frac{1}{2}$ inches) per year, because of tidal effects. The crux of the matter is what is termed angular momentum. The angular momentum of a moving body is obtained by multiplying together its mass, the square of its distance from the centre of motion, and the speed around its orbit – that is to say, the

rate of axial rotation. As we have seen, the Moon's axial rotation is the same as its orbital period (27.3 days), which is why it always keeps the same face turned toward us (all the large satellites in the Solar System behave in the same way with respect to their primary planets). Angular momentum can never be destroyed; it can only be transferred. If the rate of axial rotation is slowed down, as happened early in the Earth-Moon system, something else has to increase, and this 'something' is the distance

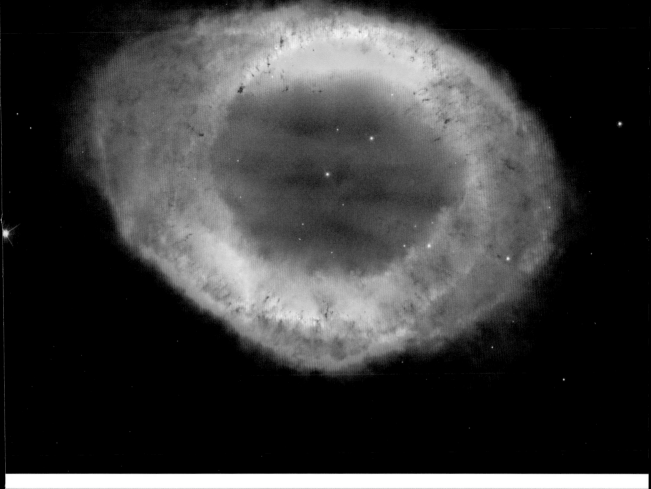

between the two bodies. The situation is rather like that of an ice skater, mid-spin. When she brings her arms in to her sides, angular momentum must be conserved and so she speeds up.

The process is not complete even now, because the Earth's rotation is still being braked by the pull of the Moon, and each day is 0.00000002 seconds longer than its predecessor, though there are also irregular fluctuations not connected with the Moon. It is these that are responsible for the occasional leap seconds that are added and subtracted from the official time. However, the Moon could not go on receding indefinitely. If it moved out to 350,000 miles it would start to draw inward again, because of tidal effects due to the Sun: its orbital period and the Earth's axial rotation period would then be equal, 47 times as long as our present day. If our world survives the Sun's red giant stage this may actually happen, but of course not until long after all life has vanished from the Earth.

▲ Butterfly Nebula (M2–9)

If this nebula is seen sliced down the middle, it is known as the 'Twin Jet Nebula', which would be highly appropriate because the velocity of the gas has been measured at 200 miles (320 km) per second!

▶ Rotten Egg Nebula (OH231)

This image gives us an insight into the fate of our Sun, since we see the planetary nebula being formed. Gas moving at millions of miles per hour rams into the surrounding gas – a supersonic shock front where the gas glows blue. It should develop into a full bipolar planetary nebula like the one above over the next 1,000 years. Why Rotten Egg? Much sulphur has been detected, and rotten eggs smell of sulphur.

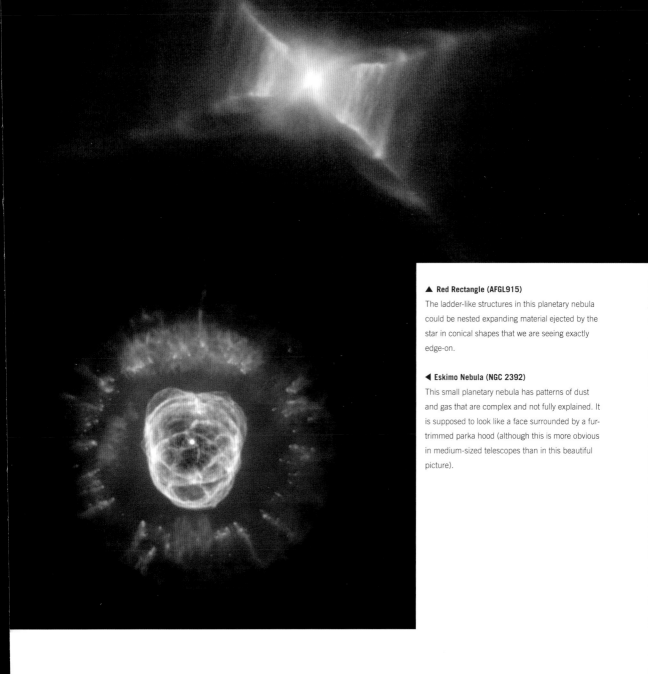

▲ Red Rectangle (AFGL915)

The ladder-like structures in this planetary nebula could be nested expanding material ejected by the star in conical shapes that we are seeing exactly edge-on.

◀ Eskimo Nebula (NGC 2392)

This small planetary nebula has patterns of dust and gas that are complex and not fully explained. It is supposed to look like a face surrounded by a fur-trimmed parka hood (although this is more obvious in medium-sized telescopes than in this beautiful picture).

▶ **An anomalous supernova?**

When supernova 1987a blew up, according to
the theory explained here, a neutron star or black
hole ought to have been left at the centre of the
expanding ring. As yet, no evidence has been
found of either.

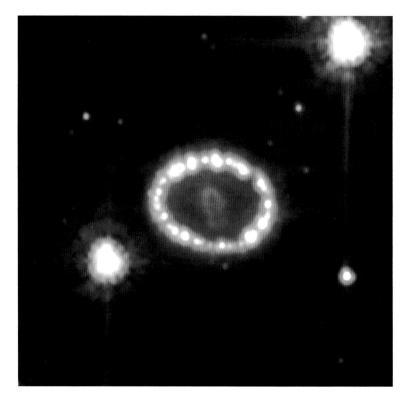

▼ **Bright white dwarf**

Also known as the Dog Star, Sirius is the brightest
star in northern skies. Telescopes reveal that it is in
fact two stars. In optical telescopes the brightest star
is Sirius A, while Sirius B, a white dwarf is 10,000
times dimmer. However, when seen in X-rays, as
below, the situation is reversed, and the white dwarf
is the strong emitter.

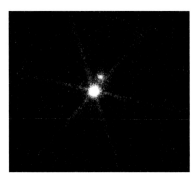

it seems that the most common is an hourglass shape, with most material being directed
along the axes of the star's magnetic field. According to this model, the planetary nebula
appears either as an hourglass or as a ring, depending on whether we are looking at it
edge- or face-on. This picture seems to be accurate on the broadest scales, but much
of the detail is more difficult to account for. Planetary nebulae are among the most
chemically interesting regions of our Universe, with many molecules being produced
by reactions driven by the light of the central star in the early stages of the nebula's
formation.

White dwarf – the bankrupt Sun

At the same time, back at the central star, now that the fuel available is exhausted, there
will no longer be anything to prevent our star collapsing under its own gravity, and this
collapse will proceed rapidly. Eventually, the density will become so great that a new
resisting force, degeneracy pressure, will begin to work against gravity. Degeneracy
pressure is a consequence of the exclusion principle, a fundamental axiom of the theory
of quantum mechanics, which holds that no two particles can ever be in the same state
– that is to say, if two particles with identical charge, mass and energy come too close
together then they will start to repel each other. The star will shrink until degeneracy
pressure exactly balances the crushing force of gravity, and at this point the collapse

ceases. The new stable state is an incredibly dense core no larger than the Earth, known as a white dwarf. A single teaspoonful of white dwarf material would weigh several tonnes. By now the Earth will have withdrawn to a distance of 170 million miles (270 million km) from the exhausted feeble remnant of the Sun.

What lies ahead? The answer must be 'very little'. The white dwarf is bankrupt; it has no energy reserve, and all it can do is to shine very dimly as it cools, eventually reaching the ambient temperature. The transition to a cold, inert, dead black dwarf takes an unimaginably long time – in fact, it may be that the Universe is not yet old enough for any black dwarfs to have been formed – it seems likely that our Sun will end its life as a tiny, dead star still orbited by the ghosts of its remaining planets.

In fact, studies of many white dwarfs have shown us the remnants of former planetary systems. The strong gravitational pull of a white dwarf should mean that its atmosphere will become what is known as fully 'stratified', in that heavier elements will sink to the bottom, meaning that the white dwarf's structure has multiple layers. Yet around a quarter of white dwarfs show a significant amount of metals – heavier elements – high in their atmosphere. Such material will settle to the bottom within a few weeks at most, and so such metals must be freshly replenished.

Where can they have come from? The source is believed to be a disk of debris, produced via the tidal disruption of the planets that once have orbited the star. Worlds which formed in a protoplanetary disk end their lives as rubble, and the composition of the material tells us about the types of planets that used to exist around stars that become white dwarfs.

Amazingly, in some systems, the transits of clumps of material within the debris disk across the face of the white dwarf have been measured. It is also possible that, in rare cases, a planet might even survive the death of its star. One such planetary survivor has been discovered by the Transiting Exoplanet Survey Satellite (TESS), in an orbit around a cool white dwarf that just happens to cross the face of the star as seen from Earth. The planet is on a tight orbit around the star, completing one orbit every 1.4 days; it is believed to have been pulled into this close orbit around the star following its transformation into a white dwarf. White dwarfs are small, and so this is a planet eleven times larger than its star; each transit causes the system to dim by a factor of roughly a half. Still, even if it appears strange to us now, maybe this solitary planet provides hope; one of the Sun's retinue may survive the end of its life.

Neutron stars and black holes

Larger stars meet a different fate. In particular, when the star is so large that the core forming a white dwarf has a mass greater than the so-called Chandrasekhar mass, 1.4 times the mass of the Sun, even the quantum effect of degeneracy pressure is not sufficient to halt the collapse. Instead, the pressure is so great that individual protons and electrons do not survive. Forced to combine, they form neutrons, and we are left with what is known as a neutron star, the density of which far exceeds even that of a white dwarf – a single sugar cube of neutron star material would weigh the same as all of humanity! Neutron stars are extremely small, no more than 15 miles across, but on average are one and a half times as massive as the Sun. If you could stand on the surface of a neutron star your weight would be of the order of 10 billion tonnes. The neutron star is actually the most common form of supernova remnant. We observe them in the guise of mysterious objects called pulsars.

▼ **Seven times dwarf**
Jupiter-sized exoplanet WD-1856b is nearly seven times the size of the white dwarf it orbits. To create this artist's impression, astronomers used NASAs TESS satellite and data from the Spitzer Space Telescope.

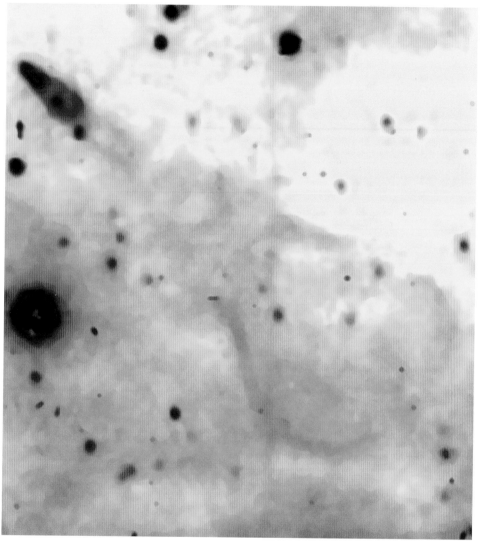

◀ Guitar Nebula

The wake left by a neutron star travelling through space at about 1000 miles (1600 km) per second created this extraordinary cosmic guitar in the interstellar medium.

In supernova events in very large stars, even a neutron star is not the end of the line in the rapid shrinking of the core. Once all its nuclear reserves have been used up, the collapse starts, but this time it is so catastrophic that nothing can stop it. The monster star goes on shrinking and shrinking and becoming denser and denser, passing through the neutron-star stage. As this happens, the escape velocity goes up. Any star with less than about eight times the mass of the Sun will end its life as either a white dwarf or a neutron star. If the star is more massive than this, the collapse is literally unstoppable and, as we have already seen, a black hole will form.

◀◀ Barrel-shaped nebula

This beautiful supernova remnant in the Milky Way has a secret that is revealed when viewed in the X-ray part of the spectrum, as here, in false colour. The brilliant blue band may be the remains of a gamma-ray burst, one of nature's most powerful explosions of energy.

▶▶ **The Antennae (NGC 4038 & 4039)**

The cores of these two colliding galaxies are the orange blobs, and a wide band of chaotic dust stretches between them. Named for their resemblance to the antennae of an insect, the similarity is far less marked in this wonderful image from the Hubble Space Telescope than it is with a less sophisticated ground-based instrument. Eventually the two galaxies will end their cosmic dance and merge, but for now they both shine brightly. This is mostly due to a burst of star formation triggered by the collision.

▼ **Arp 299**

This pair of colliding galaxies may be the best place to look for a new supernova explosion. A super star cluster in Arp 299 saw its peak of star formation 6 to 8 million years ago, and many stars are now ending their lives in supernova explosions. Four supernovae have been seen since 1990!

Pulsars

Pulsars are rapidly spinning neutron stars, which we see as pulsating sources of radio waves, with several pulses arriving each second. We have already discussed the role of angular momentum in planet formation, and it is important here, too. As the material of the star collapses to form the neutron star, it carries its angular momentum with it, and just as the ice skater bringing his arms into his side speeds up, so the forming neutron star spins faster and faster. Once the collapse is complete, the pulsar will spin at a roughly constant rate. Many pulsars that spin thousands of times a second are now known. Most of these are young; the neutron stars will gradually slow down over time.

What causes the pulses? The emission from material around the neutron star is channelled into narrow beams near the poles of the object. As the star rotates, so these beams flash across the Earth like a lighthouse beam crosses momentarily over a ship far out at sea or a watcher on the shore. When the beam is pointing toward us, our telescopes detect a pulse.

Pulsars are the most accurate clocks in the Universe; there are occasionally glitches due to some poorly-understood processes deep in the star, but apart from these rare events and the slowing-down over long timescales, they keep perfect time. They thus provide unique laboratories for astronomers. In particular there is a rare system known as the double pulsar, about which we will have more to say later. A small number of planets, detected by small changes in the timing of the pulsations, are known around pulsars. The first of them, found in 1992 around the pulsar PSR 1257+12, were the first ever planetary mass bodies found outside of our Solar System. This pulsar has three known worlds accompanying it, and they are thought to be second-generation planets, forming in a disk of debris left over after the supernova. The other two known pulsar planets may have even more exotic origins, but the existence of these unusual worlds is a sign of how pervasive the planet formation process is; even in the violent circumstances of the birth of a pulsar, planets will form if they can.

Remember, we have been discussing the evolution of the core of the star, but something more dramatic is happening outside. As the collapse is suddenly halted, the outer envelope rebounds in a stupendous release of energy. It has become a supernova.

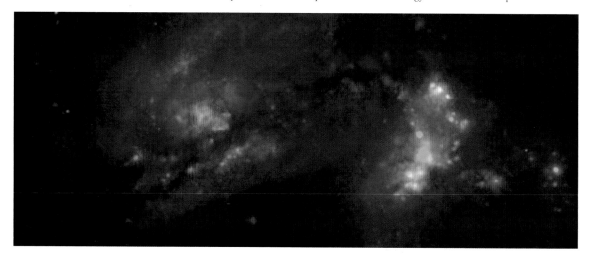

Star worlds in collision

As our Sun grows old, so too throughout the Universe old stars will die and new stars will be formed. Galaxies also are evolving and moving. Our Local Group of galaxies contains only three really major star systems; the Andromeda Spiral, the Triangulum Spiral and our own Milky Way Galaxy. Of these, Andromeda is the largest and Triangulum the smallest. Andromeda, at a distance of between two and three million light-years, is also the nearest and, under the influence of the mutual gravitational attraction between it and our Galaxy, is approaching us at a rate of 190 miles (300 km) per second. In three billion years' time, therefore, in our part of the Universe something really dramatic will occur: a collision between two large galaxies.

If a small galaxy collides with a much larger one, it is simply absorbed and will usually lose its separate identity completely; in any case, it is bound to be severely disrupted by tidal forces; its stars will be literally stripped from it every time it goes near the larger galaxy. Things are very different when two major galaxies collide.

Perhaps it is best to say at this point that although we talk about collisions between galaxies, we do not mean to imply that individual stars might collide. The space between them – remember the Sun is more than four light-years away from its nearest neighbour, Proxima Centuri – is simply too vast and stellar collisions will remain extremely rare, even in the chaotic environment of a merger between two galaxies.

The collision will take several billions of years. Andromeda will first swing past our Galaxy, and to any watchers present the tiny patch of light would become larger and larger until it came to dominate the night sky as the main interactions begin. As the reservoirs of gas in each galaxy collide, the resulting shock waves trigger the formation of many thousands of new stars, and many of these will be in brilliant clusters dominated by the hot, blue stars. The creation of many massive and therefore short-lived stars means that supernovae will be common, and the shock waves from their explosions will trigger

▲ The Mice (NGC 4676)

Three hundred million light-years away, in the constellation Coma Berenices, the cosmic capers of the pair of colliding galaxies known as the Mice will ultimately end with the pair merging into a single giant galaxy. They are nearer to completing this merger than are the Antennae, and are classified as a single system in the *New General Catalogue* (NGC).

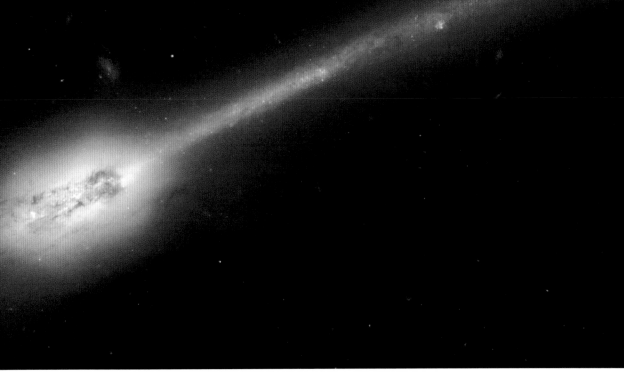

further massive bouts of star formation. The sky will be alive with bright young stars and bright star-forming regions. After swinging by, what remains of Andromeda will take perhaps 100 million years to describe a stately U-turn, before plunging headlong back into the heart of what was once the Milky Way. Much of the material will be left behind in long streamers, but over time these too will fall into the centre and a large elliptical galaxy seems the likely result. Eventually the black hole at the centre of our Galaxy may well merge with the black hole which almost certainly lies at the heart of Andromeda. It is generally believed that two black holes colliding will combine to form a single, more massive black hole. Intense radiation is bound to be released, along with what are called gravitational waves.

Gravitational waves

Two black holes colliding will combine to form a single, more massive black hole releasing what are known as gravitational waves.

Gravitational waves were predicted by Einstein's theory of relativity. They are ripples in space itself, produced by the movement of anything with mass. However, as space is stiff – it takes a big change to produce even a tiny gravitational wave – only the most energetic events, involving the rapid movement of very massive objects, produce detectable signals. After an effort lasting many decades, the first detection of gravitational waves was made by the LIGO experiment in the United States in 2015.

The experiment is simple, but it must be capable of the utmost precision. Two laser beams are sent in different directions, to rebound off mirrors placed several kilometres away, before returning to their starting point. If the two mirrors are placed exactly – and we mean exactly – the same distance away, then the two laser beams will normally return at exactly the same time, as the speed of light is constant. If a gravitational wave passes through, then it will stretch space, meaning that one of the two routes taken by the laser beam will be temporarily longer than the other. A gravitational wave interferometer like LIGO is just a device for

▶ **LIGO**

The LIGO scientific collaboration and the Virgo
collaboration made the first joint detection of
gravitational waves in August 2017 when two black
holes about 1.8 billion light-years away merged.
LIGO comprises two vast interferometers, one at
Hanford in Washington State, and the other in
Livingston, Louisiana (shown here).

comparing distances, but it does so with an accuracy greater than the size of an atomic
nucleus; that's what makes the faint rippling of space visible.

LIGO actually consists of a pair of identical experiments 3,000 kilometres (1,900 miles)
apart, one in rural Washington state and the other in Louisiana. The use of two separate
detectors helps rule out local interference, but also helps pinpoint the source in the sky of
the detected waves – rather like the way that having two ears helps your brain work out
where sound is coming from. New detectors – VIRGO just outside Pisa, in Italy, and in
India and elsewhere – are under construction to help with this task.

The first detected gravitational waves were produced by the merging of two black holes,
as are almost all of the signals seen to date. By studying the properties of the signals they
receive, which contain information about the behaviour of the black holes as they spiral
together, scientists are able to work out their mass. Through such measurements, we are
slowly understanding the population of black holes in the Universe and the result will be
a better understanding of the evolution of massive stars and the black holes they produce.

One special detection, spotted on October 16 2017, is different. This particular burst
of gravitational waves had a different pattern, and it coincided with the appearance in a
relatively nearby galaxy, NGC 4993, of a gamma-ray burst. This event is believed to be
a kilonova, the collision of two neutron stars – and it was the first event to be observed
both by normal telescopes and a gravitational wave observatory. Together, they provided a
wealth of information about this spectacular event; most striking was the demonstration
that, as the gamma-ray burst and gravitational waves were seen at the same time, gravity
moves at the speed of light, just as predicted by Einstein.

At present, our instruments can only pick up the merging of normal black holes, those
produced by the deaths of massive stars. Gravitational waves will also be produced by the

◀ **Virgo**
The Virgo interferometers two arms are each 3 km (1.86 miles) long. Located near Pisa in Italy, Virgo is part of a six nation collaboration to detect gravitational waves. Since 2007, Virgo and LIGO have shared data and jointly published results. Future instruments in India and elsewhere will add to the network's sensitivity.

grand mergers that happen after collisions like that expected between the Milky Way and the Andromeda galaxy. These lower frequency waves will need much larger instruments; so large, in fact, that we have to leave the planet. ESA is planning a constellation of three satellites flying in precise formation in the 2030s; this LISA experiment should detect the gravitational waves associated with galaxy mergers and will help us predict the future of the Milky Way's heart.

The end?

Whatever happens to the central black holes, by this time the Earth will be long gone as a habitable world, and the Sun will be nearing the last part of its career as a luminous star; it may even have already become a white dwarf. We will not be there to see it – but will anybody?

Much of the released energy will be dangerous, for example in the form of X-rays, and any life-bearing planets will be deluged with high-energy radiation that will disrupt metabolic processes and damage living tissue. The radiation may well be sufficient to wipe out even the most technically advanced civilizations. At least we may be confident that finally the activity will subside, and the newly formed galaxy will settle down. Most of the gas will have been used up in the fireworks that followed the collision, and so the rate of star formation too will have peaked. Perhaps the eventual outcome will be a system that is calm and stable, but also lifeless.

Throughout the next five billion years from now these various processes will continue – star deaths, star births, supernovae and collisions between galaxies. The most significant long-term change will be the increasing distances between the clusters of galaxies. We are drifting slowly but inexorably into the long twilight of the Universe.

NGC 4993

The elliptical galaxy NGC 4993 viewed by the HST was the first to spot a visible counterpart to the gravitational waves and gamma-ray burst, confirming they originated from the collision of two neutron stars. The event also resulted in a flare of light called a kilonova, which is visible to the upper left of the galactic centre.

18.7 BILLION YEARS A.B. (AFTER THE BANG) ONWARD

CHAPTER 7 **The End of the Universe**

▲ Graveyard of the Universe

In 10^{20} years' time, vast black holes (seen at right) will share a massively expanded Universe with the remains of stars and planets.

W hat is the ultimate fate of the Universe? It is hard to choose between a range of possibilities, but the answer must depend on the relative strengths of the only two players in the 'endgame' – gravity, and the force that drives the acceleration of the universe (called the 'cosmological constant').

Let us examine a future in which gravity wins. The expansion would come to a halt, and then reverse. Instead of observing galaxies moving away from us with redshifted spectra, we would see blueshifts as they move toward us. The temperature of the Universe would rise, and collisions between clusters of galaxies would become increasingly common. The sky would brighten, and eventually the entire Universe would end in what has been called the 'Big Crunch', something like the Big Bang in reverse.

What happens then? Perhaps the Universe could rebound so that our Universe's Big Crunch becomes the Big Bang of the next one, and so on into infinity. This cosmic recycling allows us to escape from having to suppose a moment of creation in which time began – and there is something comforting about that idea.

Unfortunately, current evidence offers little chance that the Big Crunch will ever happen; there simply is too little matter in the Universe (even including dark matter) to reverse the expansion. Gravity is not strong enough. The presence of the second player, the cosmological constant, only makes things worse, and it seems the Universe will expand for ever at an ever-increasing rate. As what happens next depends on the strength of the 'cosmological constant', this is a good time to ask whether it is indeed a constant. As yet, we simply do not have the evidence to decide. Currently, everything we know is consistent with it having a constant strength, so let us assume this, and see what happens.

Expanding forever

Long after our Sun has become cold and dead, stars will shine on as the distances between the clusters of galaxies go on increasing. It is thought that over the relatively short distances between members of these clusters, gravity will remain dominant, and strong enough to keep them together. But, over the huge distances between groups, the cosmological repulsion force brings about an ever-increasing gulf. Galaxies as viewed from each other will become very dim, and even within clusters there will be changes. As time goes by, the brilliant stars will explode to leave feeble remnants, and there will be increasing numbers of black holes. With less and less matter available to form fresh stars, there will be at first a gradual and then an accelerating descent into darkness.

From perhaps 10^{13} years hence the stars will have ceased to radiate; there will be no nuclear reserves left. Gravitational effects continue to operate, and there will be many close encounters between black dwarf stars. A star moving round the centre of a galaxy will lose energy by radiating gravitational waves, and will slowly migrate toward the galactic centre; it will join others, and the result will be the formation of supermassive black holes. It may be that the same basic principles apply to the members of what used to be a supercluster of galaxies, such as that which today includes the Local Group and also the Virgo Cluster, with all the matter collecting at the centre.

After about 10^{20} years, which is ten billion times longer than the present age of the Universe, the scene will indeed be dismal; dead stars, the ghosts of planets, vast black holes, and elementary particles and photons spread out. The whole of space will have increased to a scale beyond our comprehension; the black holes will be separated by a distance at least a hundred times greater than the present extent of the observable

▶▶ Supermassive black hole (NGC 1097)

This image shows in detail the channelling process by which matter is swallowed by the supermassive black hole at the centre of the spiral ring. More than 300 star-forming regions are visible as white spots along the ring of dust and gas.

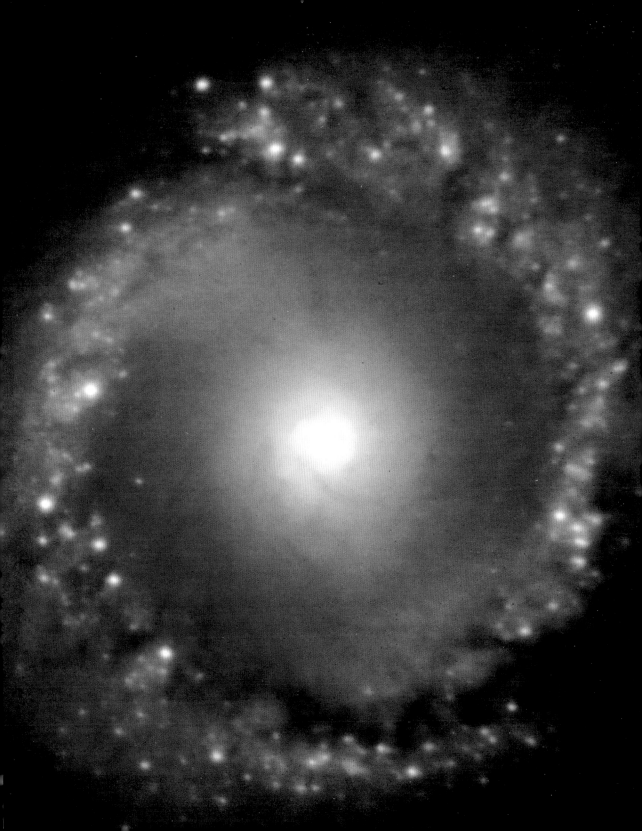

▲ The death of a black hole

This is a visualization of a black hole shrinking through emission of Hawking radiation. It is believed that ultimately all black holes will end their lives in a burst of radiation. As it shrinks, so the radiation which it emits gradually shifts toward the blue end of the spectrum.

Universe. Life everywhere will have become extinct. The Universe will not yet be dead, but the game is nearly over.

Nothing is forever

Even the black holes may not be permanent. We mentioned earlier that the vacuum of any volume of space is thought to be filled with what are termed virtual particles, which have lifetimes so short that they cannot usually turn into ordinary matter. These particles appear in pairs that are identical in every way except they carry opposite charges. The pairs quickly annihilate each other.

However, suppose a particle and its antiparticle appear just outside the event horizon of a black hole; remember that the event horizon is the boundary of the region from within which there is no escape. Before the pair can annihilate each other, as would ordinarily happen, one member of the pair may be sucked across the event horizon, while the other is ejected in the opposite direction. To an observer outside the black hole, this is tantamount to saying that the black hole has emitted a particle from within the event horizon, so that in effect the mass of the black hole decreases by the same amount as the mass of the emitted particle; the radius of the event horizon also shrinks. This can happen time and time again. The black hole becomes smaller and smaller with the emission of what is called Hawking radiation, and finally it evaporates in a final burst of radiation.

Then the ultimate: proton decay. A proton is thought to be made up of particles called quarks, but it may eventually disintegrate into lighter particles plus radiation. It may first decay into a positron (an antielectron) and a particle called a pion, which is so unstable that it would promptly decay into photons. The average lifetime of a proton is estimated to be of the order of at least 10^{31} years, so that it is hardly surprising that no instances of proton decay have been found as yet – the universe is only 10^{10} years old. But if this scenario is right, then in 10^{33} years hence there will be nothing left but a sea of photons and elementary particles.

The expansion of space will lead to an unbelievable dilution. And it has been estimated that in 10^{66} years' time the average distance between typical electrons will be over a hundred thousand times the radius of the Universe we can examine today. A googol (10^{100}) years will pass; by perhaps 10^{116} years the remaining particles will decay into

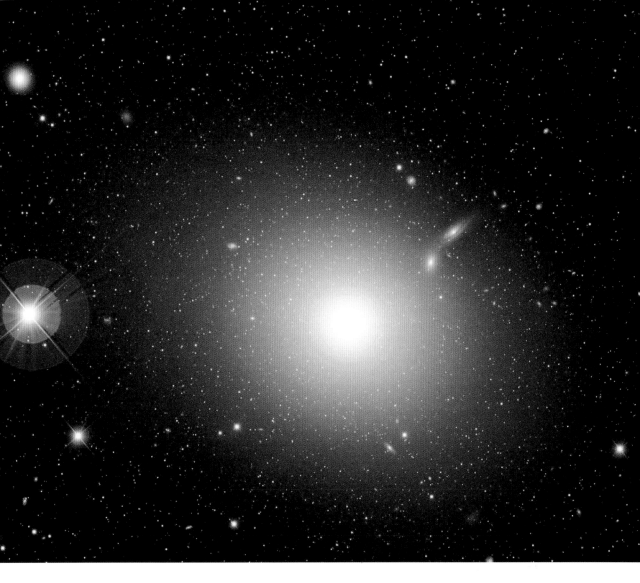

radiation. The Universe has become steadily darker and cooler, and nothing more is ever likely to happen.

Remember that this description is based on the assumption that the accelerating force remains constant in strength. What if it doesn't? It seems that if it weakens – or even if it stops acting completely – we still reach, more slowly but just as inevitably, the dismal, isolated future we have just described. But if the cosmological constant were to increase in strength, a more dramatic end would await us.

The Big Rip

At first there will be little noticeable difference. Things may happen more quickly, but we will soon be left with an isolated cluster of galaxies, which until now has always been held together by gravity. Gravity is strongest when objects are close together – on small scales – whereas the accelerating force increases in strength with separation. Eventually, though,

▲ Black holes – the last survivors

Galaxy M87 is a prominent member of the Virgo cluster of galaxies, 60 million light-years away. It has 14,000 globular clusters and spectacular jets, which extend over 8000 light-years from its centre. It radiates powerfully across the spectrum from X-rays to radio. Powering all this is the massive black hole that was imaged by the Event Horizon Telescope (see page 69).

the increasing strength of the cosmological constant will dominate on smaller and smaller scales. First the clusters of galaxies will be torn apart, leaving just an isolated galaxy in the centre of the observable universe. Structure in the Universe has, at this stage, less than a billion years to go. With sixty million years to go before the end, the individual galaxies will be torn apart, sending stars – or at least the remnants that remain – flying in all directions. The Universe is now more empty, and matter more isolated, than it is possible to imagine, but this Universe has more drama to come, in the form of what has become known as the Big Rip.

As the Universe's expansion continues, getting ever faster, eventually the matter making up the stars themselves will be ripped apart – any planets still surviving will be destroyed just thirty minutes before the end. We will be left with a sea of atoms. Even this is not the end, as if the expansion continues to accelerate the atoms themselves may be pulled apart, leaving only radiation. Even the forces that hold the atomic nucleus together can no longer resist the repulsive force, and the Universe is left as a sea of radiation and particles, much as it was just after the Big Bang but almost infinitely less dense.

This is sober science – and yet one is bound to have an instinctive, rather eerie feeling that something is wrong. A Universe that ends in any of these ways seems pointless, and there may well be a vital factor that we are missing. If at last no more events can take place, then we have nothing to measure and we may as well say that time has ended. If time ends, then we cannot speculate as to what would happen after that, because there would be no 'after'. It is hard to credit that this immensely complex and ordered universe can end in featureless chaos. Science can take us no further, and unless our powers of intellect can develop sufficiently to give us a new insight, we can do no more.

Parallel universes

At least we know that our Earth, and life upon it, has a limited future. The Universe has a more extended future, but if modern theories are correct this future is not indefinite. Therefore, does this mean that there will come a time when all intelligence ceases? We know a great deal about our Universe, but there is also the concept of 'parallel universes', coexisting with ours but in a different dimension that makes contact impossible. Such a universe may differ from ours in every way; differently made up, with a different origin, and with a different timescale. If parallel universes exist, will they, too, face extinction?

Assuming they exist, and of this we do not have the slightest proof, parallel universes may endure long after ours has either died or vanished, and if they support intelligent life, then the picture of final, complete inertness may not be valid. The trouble here is that at present there seems no way of finding out. If we do come to a final state of total inactivity, then it may be said that the whole 'experiment of the universe or universes' may have been futile, and this is something that many people will find hard to accept.

The end of the story

We have done our best to trace the history of the Universe as it is currently understood by astronomers. We began with the Big Bang, and we have journeyed through the eras of inflation, transparency, first stars, galaxies, planets and life; and further, into the literally dim and distant future of billions of years' hence. At the time of writing, this story is still the most convincing one we have, but will it still be accurate in a hundred years' time, or even a decade? Well, we just don't know!

▶▶ **The Big Rip**

Even the forces that hold the atomic nuclei together can no longer resist the repulsive force and the Universe is left as a sea of radiation and elementary particles, much as it was just after the Big Bang, but almost infinitely less dense.

EPILOGUE

◀◀ Our home in space

'We are stardust – we are golden…' (Joni Mitchell, 1969)

Just another picture of the crescent Moon? Look again…. This is our own blue-green planet seen from space. All human history happened here.

Every day, with the ever more powerful instruments at their disposal, and with penetrating theoretical work reinforced by sophisticated computer models, astronomers are learning more about the exquisitely complex Universe we live in. The more data are gathered, the more it seems that every type of event that followed the Big Bang was a necessary part of creating a blue planet that could make possible the evolution of the human race, and all the other species of animals and plants that have been born alongside us. This does not mean that we are any more important, of course, than any of our fellow creatures, or that our time is more significant than any other chapter in the unfolding of the story of the Universe.

We do hope that, in telling this story continuously in its correct order, rather than in the order in which the discoveries were made, we have communicated to the reader a feeling of the extraordinary power of the thread that runs through the evolution of the Universe – a thread into which we are all inextricably tied. Everything that we are, and everything we know, was there in that first Big Bang.

As far as we know, only on Earth is there life capable of attempting to appreciate the beauty and wonder of the Universe, from views of distant galaxies to the twinkling of a single star. This beauty can only be enhanced by our increased understanding, and in the fifteen years since the first edition of this book was written, we have been witness to an explosion of new ideas, and new information about the history of our Universe.

The discovery of exoplanets in particular has rewritten our ideas about planet formation, changed our understanding of our solar system and galvanised the search for life elsewhere. All this has happened alongside progress in cosmology, galaxy formation, stellar astrophysics and more. Will the next fifteen years be as transformative? We are optimistic, and hope that the journey through time presented in Bang!! has many more surprises in store.

Brian May, Chris Lintott and Hannah Wakeford
May 2021

PRACTICAL ASTRONOMY

How much of this amazing Universe can you see personally? The answer is: a great deal. Astronomy is just about the only science where amateurs and professionals work closely together, and where amateurs can make really valuable contributions, as well as giving themselves endless enjoyment. And to become a practical observer, you need not spend large sums of money upon expensive equipment. George Alcock, the most famous comet-hunter and nova-hunter, never used a telescope in his life. All his work was carried out with powerful, specially-mounted binoculars, which he took out into his garden!

How to become an astronomer

The following suggestions may be useful, though of course different people will have different ways of doing things, and a great deal depends upon personal circumstances and environment. But we give them as what we hope are useful tips.

1. Read some introductory books, and make sure that you know the main essentials. Having read through the previous pages you will certainly know the facts already, but in case of possible confusion, our glossary will help.

2. Join an astronomical society. There are national societies, such as the British Astronomical Association or the American Association of Variable Star Observers – all you need is enthusiasm. Moreover, most large towns and cities all over the world have local societies. By joining, you will make many new friends, and there are always people to give help and advice. In general, astronomers are sociable folk!

◄◄ **The Pleiades Cluster**
Also known as the Seven Sisters or M45. The Pleiades are an open star cluster in Taurus.

◄ **Observing the transit of Venus, 2004**
This rare event was enjoyed by a large party of astronomers at Patrick's observatory in West Sussex, England.

3. Obtain an outline star map like those on pages 164–8, go outdoors when the sky is dark and clear, and learn your way around the constellations. This is not nearly as difficult as might be thought. Remember, you can never see more than around 3000 stars at any one time without using optics, and the patterns are distinctive. Moreover, the stars stay in the same positions for year after year, century after century. It is only our nearest neighbours, the members of the Solar System, that wander about from one constellation to another, and even they keep strictly to the belt round the sky known as the Zodiac.

4. Obtain a pair of binoculars, and start seeking out selected objects such as red stars, clusters, nebulae or the phases of Venus, and the satellites of Jupiter. You may, of course, be lucky enough to be greeted by a comet. In 1997, for example, the magnificent comet Hale-Bopp hung in the sky for months.

5. By this time you will almost certainly have decided how deep your interest really is, and what branch of astronomy attracts you most. Upon this depends what equipment you will need. You will almost certainly require a telescope, and the situation here is much better than it used to be only a few years ago, because it is possible to buy a reasonably useful telescope for less than £100 or US$200. Of course, a working observatory equipped with a large telescope and auxiliary equipment costs a great deal of money, but this can come later. Also, remember that a telescope does not wear out; provided it is properly cared for, it will last for a lifetime and more with only straightforward maintenance.

We will say more about telescopes later. Meantime, let us put ourselves in the position of enthusiasts who have read this book and are now anxious to see some of the objects about which they have been reading. Also, let us assume that they are the proud owners of good binoculars plus an inexpensive telescope – say a 3-inch or 80-millimetre refractor. What is the starting point?

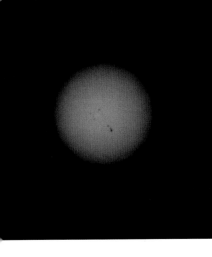

▼ **Sunspots**
Photographed by John Fletcher from Patrick's observatory.

The Sun

Learning the constellations is important, but perhaps most people will feel like starting with our nearest neighbours, the members of the Solar System. First on the list is the Sun – but bear one thing in mind from the outset: the Sun is dangerous. Staring straight at it, even when it is low down and looks quite harmless, is emphatically not to be recommended, because it emits radiation all over the electromagnetic spectrum, not only in the visible range, and obviously it is hot. To look directly at it with any telescope, or even binoculars, will cause permanent eye damage and possible permanent blindness unless suitable precautions are taken. Unfortunately, some small telescopes are still sold with 'sun caps', which it is said can be fixed in front of the telescope eyepiece to cut out the brilliance and the heat. These devices are ineffective and highly dangerous. They do not give full protection, and they are always liable to shatter without warning. They should never be used. There are filters that can be fitted over a refracting telescope, but do not try direct solar observation of this kind until you really know just what you are doing.

The answer is to use the telescope as a projector, and look at the Sun's image on a suitably positioned screen (see page 109). This is safe enough, and the image will show any sunspots that happen to be around. Observe daily, and you will see the spots being carried across the disk because of the Sun's rotation. This is fascinating, and remember that the Sun is a normal star, the only one close enough to be studied in real detail.

◄ **Lunar crater**
The 60-mile (96 km) diameter crater at the bottom
of the picture is named Plato.

Serious solar research involves specialized equipment beyond our present scope, but
there are plenty of books available. Certainly the solar observer need not brave the chill
of dark nights, and even from central city locations the Sun can sometimes be glimpsed!

The Moon

Come next to the Moon, which is 'safe' although it may dazzle you – the amount of heat
it sends us is too small to be a hazard. The main features are the seas (or 'maria'), the
mountains and the craters. The naked eye shows the main maria, and binoculars will
reveal many of the craters as well as the impressive mountain ranges. With a telescope,
an incredible amount of detail is available.

The maria are lava plains, some regular in outline and others less well defined. They
were once thought to be real seas, or at least sea-beds, and have retained their interesting
names: Mare Serenitatis (Sea of Serenity), Sinus Iridum (Bay of Rainbows), Oceanus
Procellarum (Ocean of Storms) and so on, but we now know that there have never been
areas of open water on the Moon. The lunar seas are bone dry, and are scarred with
craters; some of them have mountainous borders – thus the regular Mare Imbrium is
bounded in part by the Apennines and the Alps, with peaks higher than any in the Earth
ranges with the same names. Note that most of the maria form a connected system, the
main exception being the relatively small Mare Crisium (Sea of Crises). The maria do not
extend over the Moon's limb, and this leads on to a very important point.

The Moon's orbital period is 27.3 days. As we have seen, its axial rotation period is
exactly the same, so that the Moon's rotation is 'captured' or 'synchronous'. There is no
mystery about this – tidal friction over the millennia has been responsible, and all major
planetary satellites behave in the same way. The Moon keeps the same hemisphere turned
Earthward all the time, and before the Space Age we had no direct information about the

▲ **Lunar eclipse**

The colour of the Moon during a total eclipse
depends on the state of the Earth's atmosphere.
After a major volcanic eruption, like Krakatoa in
1883, the Moon looks very dark indeed.

averted regions, which are always turned away from us and which we can therefore never
see. Note that, as we have said earlier, although the Moon keeps the same face turned
toward the Earth, it does not keep the same face permanently turned toward the Sun, so
that day and night conditions are the same for both hemispheres—there is no 'dark side'.

However, again as mentioned earlier, things are complicated by effects known as
librations. The Moon's orbit is not circular, but markedly elliptical; following Kepler's
laws the Moon moves quickest when closest to us (perigee) and slowest when furthest
out (apogee). Yet its rate of axial rotation remains constant. This means that during each
circuit, its position in orbit and the amount of rotation become slightly out of step. The
Moon seems to oscillate very slowly and very slightly; we can peer a little way round, first
one mean limb, then round the other. This is called libration in longitude. It, together with
less important librations, means that all in all we can examine 59 per cent of the total
surface, though of course no more than 50 per cent at any one time.

The remaining 41 per cent was unexplored until 1959, when the Russians sent their
uncrewed probe Lunik 3 on a round trip and obtained the first pictures of the hitherto
unknown regions. Not surprisingly, these areas turned out to be very like the areas we
have already known, with mountains, valleys and craters. One vast sea, the Mare Orientale
(Eastern Sea) lies almost wholly on the far side, though a tiny portion can be observed from
Earth at maximum libration, and was first noted in 1948 by a certain astronomer named
Patrick Moore! American observers rediscovered it around ten years later.

A crater is at its most prominent when at or near the terminator (the boundary between
the sunlit and dark hemispheres), as its floor will be wholly or partly filled with shadow.
Many large craters have high central peaks, which catch the Sun's light while the main
floor is still dark. As the Sun rises, the shadows shrink, and even a large crater may
become difficult to identify under high illumination unless it has an exceptionally dark
floor or exceptionally bright walls.

For the beginner, full Moon is the very worst time to begin observing; there are virtually
no shadows, and the scene is dominated by bright rays issuing from a few craters, such as
Copernicus in the Oceanus Procellarum and Tycho in the southern uplands. The rays are
surface deposits, not seen under low-light conditions. In case you are wondering about
the names of the craters, they honour personalities including past observers of the Moon.
The system was due originally to the Jesuit astronomer Riccioli, who drew a lunar map
in 1651, and has been extended since, though some unusual people have found their
way there; Julius Caesar has a large crater, not for his military prowess but because of his
association with calendar reform. One crater is called Hell, but is not particularly deep; it
honours the 18th-century Hungarian astronomer, Maximilian Hell!

Other lunar features include ridges, isolated peaks, low mounds or domes, and the
crack-like rills (also known as rilles or clefts). Because the scene changes so quickly with
the conditions of illumination, it is wise to take several formations and sketch them under
different lighting, using an outline map. Persevere and it will not take you long to find
your way around the surface of the Moon. When beginning an observing session, it is
sensible to start out by scanning the area to be studied; by all means use a photographic
outline (taking full account of the main features). Do not attempt to draw too large an
area at any one time, and do not use too high a power; if the image becomes hazy or
unsteady, change at once to a lower magnification. Electronic equipment used with a
modest telescope can produce superb pictures.

Eclipses

When the Moon passes into the cone of shadow cast by the Earth its supply of direct sunlight is cut off, and the Moon turns a dim, often coppery colour until it passes out of the shadows again. It does not vanish completely, because some rays of sunlight are bent on to it by way of the shell of atmosphere surrounding the Earth. Lunar eclipses may be either total or partial – and obviously they can only happen at full Moon! They are not important, but they are lovely to watch and excellent to photograph.

The planets

The planets have always been favourite targets for owners of small or medium-sized telescopes, and until less than a century ago it is probably true to say that most of our knowledge of their surface detail was due to amateurs. This is no longer true, but planetary observations are as fascinating as ever, partly because one never knows what will happen next.

   The inner members of the Sun's family, Mercury and Venus, are not wildly exciting. Mercury is not likely to be seen at all except if a deliberate search is being made, because it always stays close to the Sun in the sky; with the naked eye it can only be seen when at its best, either low in the west after sunset or low in the east before dawn. The best views are obtained during daylight, when Mercury is high up – but so is the Sun. To locate Mercury you need a motorized telescope with accurate positioning equipment. Sweeping around for the planet is most unwise; sooner or later the hapless observer will look at the Sun by mistake. Even when Mercury is found, all that will be seen will be the characteristic phase. Only space vehicles have so far shown the Mercurian craters, plains and mountains.

   Telescopically, Venus is only slightly more rewarding. It is much brighter than any other star or planet, and really keen-sighted people can see it with the naked eye even in broad daylight. The phase is very evident, but generally the disk will appear almost featureless, with no markings apart from vague cloudy patches and bright areas. The dense atmosphere hides the surface; ordinary telescopes will not penetrate it; there is no such thing as a sunny day on Venus. Space research methods have had to be used to reveal the craters, the lava-flows and the volcanoes, and we now know that although named after the goddess of love, Venus is a hostile place. There is little scope for the visual observer.

Transits

There are occasions when Mercury and Venus pass in transit across the face of the Sun. Mercury does so reasonably often – the next occasion will be May 9, 2016 but will only be observable with optical aid. The next transit of Venus will be on June 6, 2012, after which there will be no more until December 11, 2117. Venus is quite conspicuous in transit – but of course, when observing transits all the normal precautions for solar observing must be taken.

Mars

Mars is different because the atmosphere does not hide the surface features, and when the planet is well placed, a small telescope will show the dark areas, the ochre 'deserts' and the white polar caps. Yet Mars is a small world, and is seen well for only a few weeks when the planet is in opposition – positioned on the opposite side of the Earth

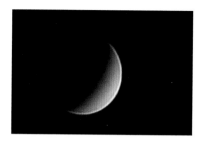

▲ **The crescent of Venus**
Some observers have reported an ashen light on the dark side, which has been attributed to everything from lightning to city lights under the clouds! The most likely explanation, perhaps sadly, is an optical illusion.

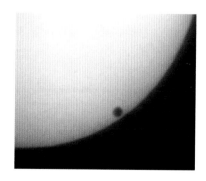

▼ **Transit of Venus**
Brian took this picture of his projection of the 2004 transit of Venus on a card at Patrick Moore's observatory in Sussex.

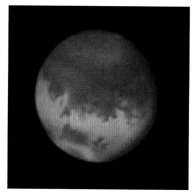

▲ Mars

The Martian polar caps wax and wane with the seasons and are cleary visible in small telescopes. The dark areas are sometimes obscured by dust storms, which start small but can engulf the entire planet.

from the Sun. Moreover, it does not bear magnification well, so that the observer has to take advantage of the best nights around the time of opposition. Not all oppositions are equally favourable; the orbit of Mars is markedly eccentric, and the best oppositions occur when the planet is at opposition and perihelion (closest point to the Sun) at the same time. In 2003, the distance to Mars was less than 35 million miles.

Away from opposition, Mars shows a distinct phase, resembling the Moon a few nights from full. When making drawings, always remember to allow for phase. The rotation period is over half an hour longer than ours, so that there is no urgent rush to position the main features, as there is with Jupiter.

If you are making a sketch, first, put in the polar cap (if visible) and the dark markings. Look round for any cloud phenomena, change to the highest available power, and add the fine detail. Record the time of observations plus the details of the telescope, magnification and weather conditions and also the longitude of the central meridian. Be careful to draw only what you can see – not what you expect to see. Patrick has vivid memories of his first view of Mars through the 24-inch refractor at the Flagstaff Observatory in Arizona, used by Percival Lowell to draw his famous canal network. Would canals be seen? It was a relief to find that they were conspicuous only by their absence.

Visual observations of Mars are being to some extent supplemented by the excellent photographs obtained by amateurs using electronic instruments in combination with small telescopes. The images shown in this chapter are better than any professional photographs available of Mars in, say, 1960. We can see the great volcanoes such as Olympus Mons – but from Earth we could never feast our eyes upon the complex summit caldera; neither could we ever see into the depths of the colossal Valles Marineris. Only spacecraft can show us these wonders.

Asteroids

Beyond the orbit of Mars we come to the Asteroid Belt. These midgets look like stars through a telescope, after all even the largest of them, Ceres, is a mere 600 miles in diameter. However, detailed times of their positions are available, and asteroids are easy to photograph. Some asteroids, such as Hermes, are classed as PHAs (Potentially Hazardous Asteroids), because their orbits cross that of the Earth. Future collisions cannot be ruled out and amateur astronomers have a role to play here. There so many PHAs that professional astronomers are hard pressed to keep track of them all.

Jupiter

Jupiter and Saturn, the giants in the Sun's family, are of special interest to the amateur observer. Jupiter, with its belts, its spots and its Galilean satellites, is always changing. Saturn's rings make it arguably the most beautiful object in the sky.

Jupiter, large enough to hold over a thousand globes the volume of the Earth, shows a yellowish, flattened disk crossed by cloud belts. Generally there are two main belts, one to either side of the Jovian equator, with others at higher latitudes. Jupiter has a gaseous surface, and does not rotate in the same way as a solid body would do. In what is termed System I (between the north edge of the South Equatorial Belt and the south edge of the North Equatorial Belt) the rotation period is 9 hours 51 minutes. Over the rest of the planet (System II) the period is five minutes longer, and individual features, such as spots, have periods of their own, so they drift around in longitude.

◀ Jupiter
The planet's brown equatorial belts are clearly visible.

The belts, due to droplets of liquid ammonia, dominate the scene. Spots can become more prominent but usually only for a limited period. The exception is the Great Red Spot, which has been seen ever since the 17th century; it is brick-red, and over 20,000 miles (30,000 km) long. It can vanish for a while but always returns. There are some well-marked white spots, but these are always temporary; in 2006 a smaller red spot was seen.

Jupiter is a quick spinner, so that one cannot afford to linger when making a sketch; the main drawing should be finished in less than 10 minutes, with finer details added afterwards as quickly as possible. Rotation periods of discrete features can be found observationally. Time the moment when the feature crosses the central meridian, and then use tables to work out the longitude. This is not difficult – the central meridian is easy to find because the planet's disk is so flattened. With practice, estimates can be made to an accuracy of within a minute (sixtieth of a degree). Work of this kind used to be very valuable. We have to admit that this is no longer so because of the improvements in imaging. But Jupiter remains fascinating, and quite apart from the surface details watch too for the transits, occultations and eclipses of the Galilean satellites.

For sketches, it is sensible to use prepared blanks. Make sure to look for anything unusual. In 1994 the fragments of Comet Shoemaker-Levy 9 cascaded into Jupiter, causing scars that persisted for months. Any small telescope was capable of showing them when they were first formed.

Saturn and beyond

Saturn's rings are able to provide endless enjoyment. It is interesting to compare the amateur photograph overleaf taken with an 15–inch telescope with a view from the Hubble Space Telescope. You have to look carefully to decide which is which. Occasionally white spots appear on the disk. One was discovered in 1933 by Will Hay, the stage and screen comedian. Another spot of the same kind was discovered in 1990 by an American amateur, Stuart Wilber. Saturn is a less active world than Jupiter but it is still capable of springing surprises.

Photographing the outer giants, Uranus and Neptune, is very easy, though of course no surface details can be seen. There are also the bodies of the Kuiper Belt, of which Pluto is the brightest though not the largest. Imaging these objects is really useful in helping to keep track of them.

▶ **Saturn's rings**

Saturn's beautiful rings are not always visible in
small telescopes. This view was created from a stack
of 9000 separate images.

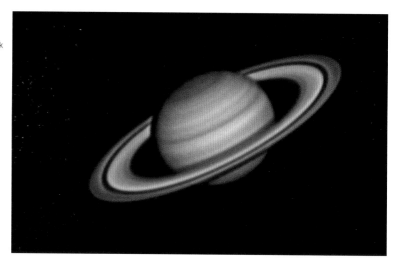

Finally, in the Solar System, we come to the most erratic members of the family; comets
and meteors. There are always various comets within range of modest telescopes, though
really spectacular visitors, such as Hyakutake and Hale-Bopp are depressingly rare. Some
comets look rather like tiny patches of shining cotton-wool, and to locate them you will
need a detailed star atlas and a telescope fitted with accurate setting circles. Of course
really spectacular comets, such as Hale-Bopp, are well seen in binoculars.

▶ **Comet Hale-Bopp**

Photographed by Patrick in 1997, this lovely comet
will not return for 2360 years. Note the proximity to
the chimney stack of Patrick's house.

Meteors can be seen at any time but the annual showers are interesting to watch; good photographs can be taken with an ordinary camera. The August Perseids are always reliable; given a clear dark sky for a few minutes between around August 9 and 18 you will be very unlucky not to see several Perseids.

The stars of the northern hemisphere

Come now to the stars. This is no place to give a full review of the constellations so we have given just a few charts, which will, we hope, enable the newcomer find his way around. For northern-hemisphere observers there are three key groups, Ursa Major, Orion and Pegasus, so let us start there.

Ursa Major, the Great Bear, is circumpolar from the latitudes of London or New York – that is to say, it never sets, so that it can always be seen somewhere or other

▼ Zodiacal light

The evening zodiacal light in Gemini photographed in Tenerife in 1971. It was taken by Brian, with a Pentax 35mm film camera on a tripod. As the Earth rotated during a time exposure of a few minutes, the stars left trails of light on the image. The zodiacal light is a cone of light visible in the evening or the morning. It is believed to be sunlight reflected from dust in orbit around the Sun. Brian's PhD research was concerned with determining the motions of the dust.

provided that the sky is sufficiently dark and clear. Seven of its stars make up the pattern known familiarly as the Plough or (in North America) the Big Dipper. Because they are so useful, we have named these stars in Map 1: Alkaid, Mizar, Alioth, Megrez, Phad, Merak and Dubhe; their magnitudes are between 1.7 and 2.4 except for Megrez, which is considerably fainter. Merak and Dubhe are known as the Pointers, because they show the way to Polaris, the Pole Star, in Ursa Minor (the Little Bear). Polaris is within one degree of the north celestial pole, and so seems to remain almost stationary in the sky. Ursa Minor contains only one other reasonably bright star, the orange Kocab (2.1).

On the far side of Polaris with respect to Ursa Major is Cassiopeia, whose five main stars form a W pattern (magnitudes 2 to 3). Like Ursa Major, Cassiopeia is circumpolar.

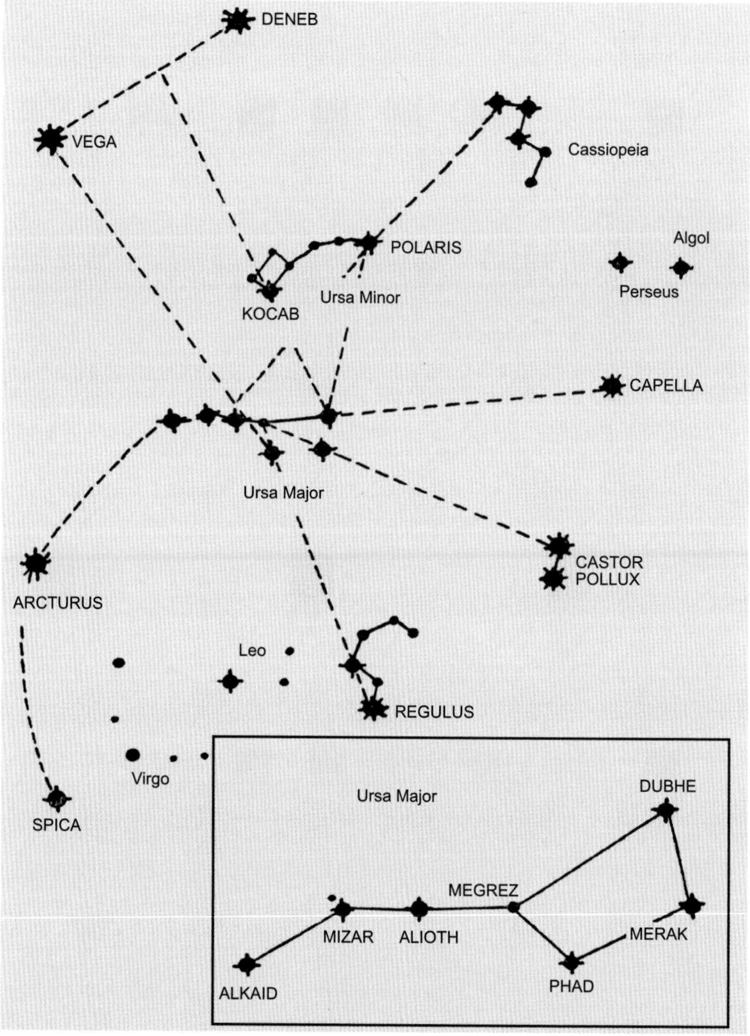

▶ **Map 1**

Ursa Major.

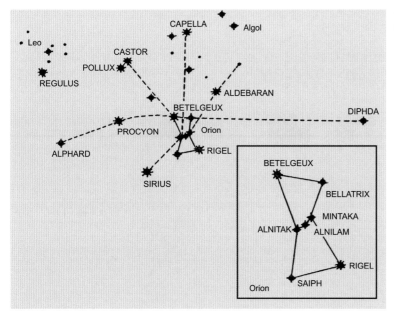

◀ **Map 2**

Orion.

When the Bear is high in the sky Cassiopeia is low down, and vice versa. From the Bear you can find the brilliant bluish-white Vega (0.1) in Lyra (the Lyre), and also Deneb (1.3) in Cygnus (the Swan). Both these are circumpolar from latitudes of Britain and the northern United States, though at their lowest they almost graze the horizon.

The remaining stars in Map 1 are not circumpolar. The 'tail' of the Bear shows the way to the orange Arcturus (-0.06) in Bootes (the Herdsman). And then to Spica (1.1) in Virgo (the Virgin). From Megrez and Merak you can locate Castor (1.5) and Pollux (1.2), the Twins (Gemini); Pollux, the brighter of the two, is orange, while Castor – a fine binary – is white. Also find Leo (the Lion), led by Regulus (1.4); extending from Regulus is a curved line of stars known as the Sickle.

Map 2 shows Orion, the celestial hunter, with his brilliant retinue. Orion dominates the evening sky in northern winter (the southern summer), and cannot possibly be mistaken. The leading stars are Rigel (0.2) and Betelgeux (variable between 0.3 and 0.8); Rigel is glittering white, while Betelgeux is the orange-red supergiant. The other main stars of Orion are around the second magnitude; Alnitak, Alnilam and Mintaka make up the belt – immediately south of that, visible with the naked eye as a misty blur, is the sword, containing the famous Orion Nebula. Southward, the belt shows the way to Sirius (-1.4) in Canis Major (the Great Dog), which is pure white, but when low down seems to flash all the colours of the rainbow! Northward, the belt points to the orange Aldebaran (0.9) in Taurus (the Bull) and on to the magnificent open cluster of the Pleiades, or Seven Sisters. Aldebaran, the 'Eye of the Bull' looks much the same colour as Betelgeux, but is not nearly so luminous – minus 140 times as powerful as the Sun. Extending from it in a V-pattern are the stars of the Hyades cluster, but Aldebaran is not a cluster member; it is only 65 light-years away, and just happens to lie between the Hyades and ourselves.

▶ **Map 3**

The Summer Triangle and adjacent constellations.

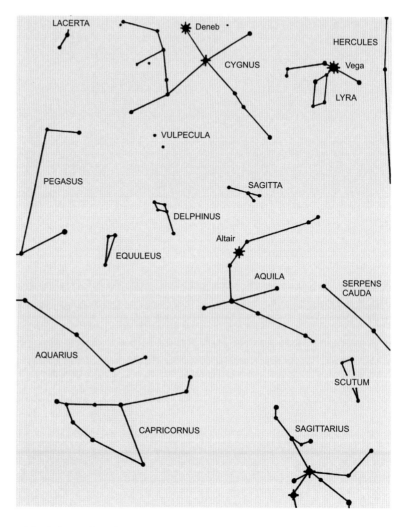

Map 3 shows what is known commonly as the Summer Triangle (an unofficial name, popularized by Patrick during a television broadcast around 1960!). It consists of Vega in Lyra, Deneb in Cygnus and Altair (0.8) in Aquila (the Eagle), and is linked to either side by fainter stars. Lower down is Sagittarius (the Archer) with the superb star-clouds in the direction of the centre of the Galaxy.

Map 4 shows Pegasus, whose four main stars make up a square; Markab (2.5), Algenib (2.9), Scheat (variable from 2.4 to 2.9) and Alpheratz (2.1). For some unknown reason Alpheratz has been given a free transfer to the adjacent constellation of Andromeda, with Mirach and Almaak (each 0.1). Here we see the great spiral galaxy M31. It is just visible with the naked eye; binoculars show it easily, though only as a misty patch. The map also shows Fomalhaut (1.2) in Piscis Australis (the Southern Fish), which is always very low

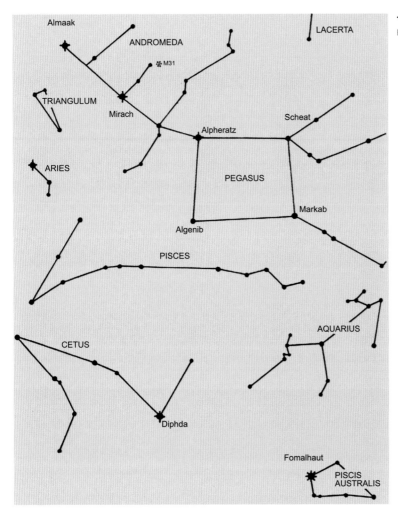

down from northern latitudes. Pisces (the Fishes), and Aquarius (the Water-bearer) are dim Zodiacal constellations; Aries (the Ram) does have one bright star, Hamal (2.0).

Sky survey

Next let us survey the sky, soon after darkness falls, for each season: Winter: Let us say mid-January at 10 pm (22.00 GMT). Ursa Major is in the southeast, Cassiopeia is in the southwest; Capella is almost overhead, with Vega very low in the north. Pegasus is setting in the west, Orion is high in the south, with Sirius outstanding and Leo rising in the east.

Spring: Mid-April, 22.00 GMT. Ursa Major is overhead; Capella is descending in the west, with Vega gaining altitude in the east. Cassiopeia is in the north. Orion has almost

▶ **Map 5**

The major constellations of the
southern hemisphere.

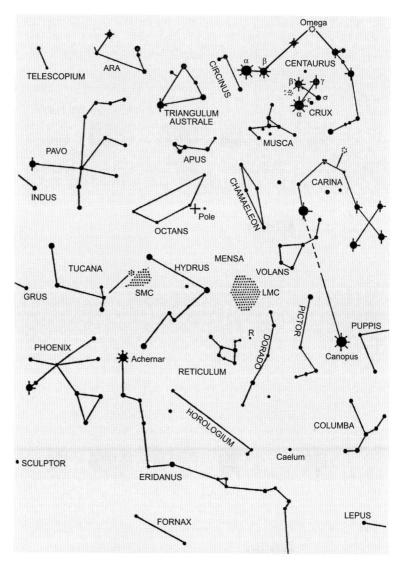

disappeared, but Sirius is still visible; Arcturus is prominent in the east, while Leo and
Virgo are high in the south.

Summer: Mid-July, 22.00 GMT. Ursa Major is in the north-west, Cassiopeia is in
the north-east; Vega is almost overhead, and Capella is near the northern horizon. The
Summer Triangle dominates the south. Arcturus is fairly high in the west, Virgo is setting;
Sagittarius is low in the south. Much of the southern aspect is occupied by large, dim
constellations – Hercules, Ophiuchus and Serpens, with only one bright star, Rasalhague
(2.0) in Ophiuchus, the Serpent-bearer.

Autumn: Mid-October at 22.00 GMT. Ursa Major at its lowest in the north, Cassiopeia is overhead. Pegasus is high in the south; Fomalhaut is low in the south; the Summer Triangle is still very prominent. Aldebaran and the Pleiades have risen; Orion will soon follow. Winter, with its frosts and snows, lies ahead!

The stars of the southern hemisphere

So much for the northern hemisphere. Now let us go south – for example, to South Africa or Australia. Remember that you do not actually have to cross the equator to see the Southern Cross; it rises from anywhere south of latitude 35 degrees north.

The Southern Cross, Crux, is the most famous of the southern constellations, shown in Map 5. Its four main stars are Acrux or Alpha Crucis (0.8), Beta Crucis (1.3), Gamma Crucis (1.6) and Delta Crucis (3.1). In fact Crux is shaped more like a kite than a cross; it is not nearly so cruciform as Cygnus in the northern sky. It is also much brighter and smaller – rather surprisingly, it is the smallest of all the 88 accepted constellations. It is almost surrounded by Centaurus (the Centaur); Alpha Centauri (-0.3) and Agena or Beta Centauri (0.6) point to it. The two Southern Pointers are not associated; Alpha, at 4.4 light-years, is the nearest of the bright stars, while Agena is a giant 530 light-years from us and 13,000 times as luminous as the Sun. In the Centaur, look for the globular cluster Omega Centauri, an easy naked-eye object.

Unfortunately, the south celestial pole is not marked by any bright star; we have to make do with Sigma Octantis, below the fifth magnitude and hidden by any mist or haze in our atmosphere. The Pole lies between Crux and Achernar (0.5) in Eridanus (the River). Also on this map is Canopus (-0.7), the brightest star in the sky apart from Sirius; it is a supergiant. It lies south of Orion; it can just be seen from Alexandria in Egypt, but not from the other great centre of ancient astronomy, Athens.

Next, a survey of the southern sky through the seasons. Summer: Observing in January in mid-evening. Orion is high in the northern part of the sky; Sirius is very visible, Canopus is almost overhead, Crux is in the south-east with the Southern Pointers. Capella is very low in the north.

Autumn: Mid-evening in April. Crux and Centaurus are very high, Canopus is sinking in the west, Sagittarius is gaining altitude in the south-east. Arcturus is in the north-east, but Achernar is low in the south.

Winter: Mid-evening in July. Scorpius and Sagittarius are almost overhead and the star clusters are glorious. Crux and Centaurus are sinking in the south-west, Achenar is rising in the south-east. Vega can be seen rather low in the north; Fomalhaut is conspicuous in the south-east.

Spring: Mid-evening in September. Crux and Canopus are low in the south; Pegasus can be seen in the north and the northern Summer Triangle, here perhaps better referred to as the Winter Triangle. Fomalhaut is almost overhead.

Finally, on no account fail to look at the two Magellanic Clouds, in the south polar area. With the naked eye they look rather like detached parts of the Milky Way, but they are in fact satellite galaxies, over 169,000 light-years away. They contain objects of all kinds – giant and dwarf stars, clusters, nebulae and novae. There have been two observed supernovae, one in 1885 and the other in 1987. Northern observers always regret that they lie in the deep south. We appreciate that this ramble round the constellations is very fragmentary, but it may act as a start. The skies are never dull.

▶▶ **La Palma**

A panoramic view of the Roque de los Muchachos observatory in La Palma, Canary Islands.

TIME LINE OF THE UNIVERSE

Time after the Big Bang (A.B.)	Event	Years ago / future
0	Big Bang	13.8 billion years ago
10^{-35} to 10^{-33} seconds	Inflationary period	
10^{-33} seconds	Birth of quarks and antiquarks. They annihilate each other, leaving a slight excess of quarks.	
10^{-5} seconds	Quarks combine to form protons and neutrons.	
10^{-3} seconds	Formation of hydrogen and helium atoms	
1 to 3 minutes	Formation of light elements up to boron	
370,000 years	Emission of CMB – Universe becomes transparent	
200 million years	Birth of first stars, reionization	13.5 billion years ago
3 billion years	Formation of mature galaxies, quasars, and the oldest stars in Milky Way	10.4 billion years ago?
9.1 billion years	Our Solar System including Earth formed	5.6 billion years ago
9.9 billion years	First fossils formed	3.8 billion years ago
13.4 billion years	First reptiles	320 million years ago
13.5 billion years	Africa splits from America; the dinosaurs appear	200 million years ago
13.64 billion years	End of the dinosaurs; small mammals diversify	65 million years ago
13.695 billion years	Primates including the first apes evolve.	5 million years ago
13.6998 billion years	Homo sapiens.	195,000 years ago
13.6999 billion years	End of last ice age, dawn of modern world	10,000 years ago
13.8 billion years	**Present day**	
14.7 billion years	Earth becomes uninhabitable.	1 billion years hence
18.7 billion years	Sun a red giant, destruction of Earth	5 billion years hence
23.7 billion years	Sun becomes a white dwarf	10 billion years hence
10^{14} years	Galaxy- and star-formation cease	A hundred thousand billion years hence
10^{36} years	50 per cent of all protons have decayed.	
10^{40} years	All protons gone, black holes dominate	
10^{100} years	Black holes disintegrate	
10^{150} years	Photon age: Universe reaches extreme low-energy state?	

GLOSSARY

Atom The ancient Greeks believed that matter could be broken down into indivisible units, which they called atoms. The modern idea of an atom is remarkably similar; it has a nucleus made up of positively charged particles called protons and neutral neutrons, which is surrounded by low-mass, negatively charged electrons. Electrons and protons have the same charge, although with opposite signs, and so a neutral atom must have an equal number of both. Carbon has 12, for example, while an atom of the lightest element – hydrogen – consists of a solitary proton and a single electron. Classically, physicists thought of the electron orbiting the nucleus just as the planets orbit the Sun, but in quantum physics things are much less straightforward.

Antimatter Modern theories of particle physics predict that every type of particle has an antiparticle, which has the opposite electrical charge but is otherwise identical. Collectively these antiparticles are known as antimatter. For example, the electron's antiparticle is the positron. When particle and antiparticle collide, they annihilate, releasing energy and (at least in science fiction) powering spaceships. In the very early stages of the Universe's evolution there were equal amounts of matter and antimatter, and it is unclear how we ended up in a Universe made overwhelmingly of matter.

Baryon A baryon is a particle that is composed of quarks. Examples include quarks themselves, neutrons and protons. Astronomers use the term 'baryonic matter' to distinguish the ordinary material in the Universe from the mysterious dark matter.

Billion Throughout this book we have used the standard scientific definition of billion, which is a thousand million (1,000,000,000 or 10^9). The older, English definition according to which a billion equalled a million million is now almost obsolete.

Black body An idealized emitter and absorber of radiation, to which a star approximates quite well. Any hot body emits electromagnetic radiation, and the spectrum of this emission for a black body is entirely determined by its temperature. A plot of energy against frequency (or colour) for a black body produces a smooth 'hump-backed' shape (see p.58) with a maximum intensity that moves to a higher frequency as the temperature increases. If a metal object such as a poker is heated, it appears first red, then orange, then yellow and eventually white hot. Similarly, the hottest stars have a blueish-white colour.

Black hole A black hole is a body that has a strong enough gravitational field to prevent even light – the fastest thing in the Universe – from escaping. The radius within which a particular mass must be confined to form a black hole is called the Schwarzschild radius. Considered for many years to be theoretical curiosities, there is now strong evidence that black holes exist. It seems that most galaxies have a supermassive black hole (as massive as several million suns) at the centre.

Brown dwarf A brown dwarf is an astronomical body that is more massive than a planet (greater than ~13 Jupiter masses) but smaller than the mass required to fuse hydrogen at its core to become a star (around 80 Jupiter masses). Brown dwarfs are able to burn deuterium at their cores which makes them distinct from the less massive planets.

Charge Electric charge is a property shared by particles such as quarks, protons and electrons. Positive and negative electric charges attract each other, and it is this force that keeps the negative electrons bound to the positive nucleus in a neutral atom.

Circular polarization An electromagnetic wave of light that has a constant magnitude but is rotating (spiraling) clockwise (right-handed) or anti-clockwise (left-handed) around its direction of motion through space.

Comet An icy body, best described as a 'dirty snowball'. Comets are believed to form in the outer regions of the Solar System, where they remain in a 'reservoir' of comets known as the Oort Cloud until disturbed. Gravitational disturbances, such as the passing of a nearby star, can alter the orbit of a comet so that it swings into the inner Solar System. As it nears the Sun, the ice melts, and the familiar tail (which always points away from the Sun) forms. Although some comets are thrown out of the Solar System altogether, most become periodic and make regular visits to our neighbourhood. The first of these to be identified was Comet Halley, which last passed through the inner Solar System in 1985/86 and has a period of 76 years. Recent spectacular comets have included Hyakutake and Hale-Bopp in the late 1990s, and Comet Shoemaker-Levy 9, which crashed into Jupiter in 1994.

Constellation A group of stars that appear near each other in the sky, forming a recognizable pattern. The stars in a constellation have no physical link to each other and may be many thousands of light years apart. Although ancient cartographers chose to create their own constellations, in 1930 the International Astronomical Union selected 88 for their official list. Many of the best known groups – such as the Plough, which is merely part of Ursa Major – are not official constellations but are known as asterisms. The largest constellation is Hydra, the Sea Serpent, and the smallest is Crux, the Southern Cross. Although the constellations appear constant on the timescales of our human lifetimes, the stars are slowly moving and over time their familiar patterns will disappear.

Dark matter Over the last fifty years astronomers have come to realize that most of the matter in the Universe is composed not of ordinary atoms and molecules, but is in the form of some exotic dark matter. In fact, over 80 per cent of the mass in the Universe is made up of dark matter, which interacts with 'normal' (or baryonic) matter almost exclusively through gravity. Particle physics predicts the existence of a family of weakly interacting massive particles (or WIMPs), which may make up the dark matter, but this theoretical prediction has yet to be confirmed by either observation or experiment. Although the identity of dark matter has yet to be confirmed, observations allow scientists to constrain its properties; the majority of the evidence suggests that the correct model is one involving 'cold dark matter' – slow moving, massive particles.

Dimension A coordinate necessary to specify one's position. In everyday life we are used to three dimensions – length, width and height – along with time, which should also be considered a dimension. Some of the more exotic theories of particle physics suggest that there may be other 'hidden' dimensions, which only reveal their presence through the results of high-energy experiments.

Doppler effect The Doppler effect is most familiar from the change of pitch heard in a siren as an ambulance drives past. The waves emitted from an approaching source are compressed, and hence appear to have a higher pitch than those from a stationary source. Conversely, waves emitted from a receding source are stretched and hence appear to have a much lower pitch. The greater the relative speed between the source and the observer, the greater the shift in wavelength. The same effect applies to light; the stretching of the light means that light from a receding source will appear reddened, and that from an approaching source will appear shifted toward the blue end of the spectrum. Observations of galaxies beyond our own Milky Way show that all but the closest have spectra that are shifted toward the red end of the spectrum, revealing that they all appear to be receding from us. The further away the galaxy is, the faster it appears to be moving away from us, and it was this discovery that provided the first observational evidence for an expanding Universe.

Ecosphere The region in a Solar System within which the temperature is such that liquid water could exist on the surface of a rocky planet and therefore – assuming, of course, that all life needs water as life on Earth does – the region within which life could exist. In our own Solar System, Venus is closer to the Sun than the ecosphere and currently (although things could have been different in the past) Mars lies further away. Searches for planets around other stars are not yet sensitive enough to identify Earth-sized planets, but at least one Jupiter-mass object has been found within its system's ecosphere. Liquid water may therefore exist on any large moons orbiting that distant planet and potentially support life.

Electromagnetic radiation Visible light is just one part of the spectrum that stretches from ultra-high energy gamma rays and X-rays through the ultraviolet, then through the visible into the infrared, microwaves and then finally the radio wavelengths. Electromagnetic radiation in all of these forms is composed of electric and magnetic components, which move at the speed of light (see p.44).

Electron A low-mass particle (weighing the same as less than one thousandth [0.001] of a proton) with unit negative electric charge. Unlike protons and neutrons, electrons are not made up of quarks and appear to be truly 'fundamental' particles that cannot be broken into smaller parts.

Energy The law of conservation of energy (otherwise known as the first law of thermodynamics) is one of the most fundamental of all physical laws. It states that energy is neither created or destroyed, but can only be converted from one form to another. The famous equation $E=mc^2$ simply states that mass can be converted into energy or, equivalently, that mass is simply another form of energy. The nuclear reactions at the centres of stars convert mass into radiation and thermal energy.

Equator The imaginary circle drawn on a sphere so as to be an equal distance from both poles. We are relatively familiar with the Earth's equator, and the projection of this line on to the sky defines the celestial equator. It is useful as a point of reference for our coordinate systems, but its position on the sky has no physical significance.

Exoplanet A planet that orbits a star other than the Sun. Thousands of exoplanets have been discovered since the early 1990s. The most common size of exoplanets found are larger than the Earth but smaller than Neptune (4 x Earth's radius (Re)). These are known as super-Earths if less than 1.7Re or mini-Neptunes if they lie between 1.7Re and 4Re. These worlds likely represent more than half of the exoplanets in our Galaxy, yet we do not have one in our Solar System as an example. Other expolanets include hot Jupiters (Jupiter-sized planets orbiting their stars on less than 10-day orbits), super Jupiters (planets with masses greater than 3 Jupiter masses but less than the deuterium burning limit (see Brown Dwarfs)), and lava worlds (rocky planets so close to their stars that their entire surface is likely to be completely molten).

Galaxy From the Greek word for 'milk,' the term galaxy was first applied to the Milky Way, which was seen as a bright strip of starlight running through the sky. Once it became clear that our own Galaxy was only one of many billions, the term was applied to mean any large group of stars and other material that exists as an independent system, held together by its own gravity. The two major classes of galaxies are the ellipticals and the spirals. The ellipticals are large, spherical systems of old stars with relatively little gas remaining to be converted into stars. By contrast, a spiral is characterized by a disk containing spiral arms that mark ongoing star-forming activity surrounding a central, older bulge. It was believed for many years that elliptical systems formed from the collision of two spirals, but the process appears to be more complicated than that.

Gamma-ray burst The most powerful explosions in the Universe, these rapid events were first detected by satellites that were monitoring the Earth for signs of hidden testing of nuclear weapons during the Cold War. At least some of them are associated with extreme supernova outbursts known as hypernovae but others may be due to exotic phenomena such as the collision of black holes and neutron stars. As they are extremely luminous, they are visible even in parts of the distant Universe.

Gravity Although gravity is intrinsically the weakest of the fundamental forces, it is the only one of the traditional forces that acts on astronomical scales. Of the others, the strong and weak nuclear forces are extremely weak over large distances, and the electromagnetic force from positive and negative charges cancels out. The gravitational attraction between two objects is proportional to their masses and inversely proportional to the square of their separation. In other words, two masses moved so that the distance between them is halved will attract each other four times as strongly. The first systematic theory of gravity was due to Sir Isaac Newton, whose theories were expanded by Albert Einstein in his general theory of relativity.

Habitable zone The habitable zone is the region around a star where the Earth would remain or have been habitable if placed there. This is an addition to the temperate zone which defines the distance from a star a planet would need to orbit to have a temperature between 0 and 100 centigrade such that, if it had a surface, liquid water could exist and pool into lakes, seas, or oceans. The habitable zone considers the added existence of an atmosphere surrounding the planet either reflecting enough light to cool it (if closer to the star) or containing enough greenhouse gases (CO_2, H_2O, CH_4) to maintain a warm temperature (if it is further from the star).

Heat The scientific definition of temperature is rather different from the everyday one. The higher the temperature of a gas, the faster the atoms that compose it are moving. By contrast, 'heat' is usually used to mean the quantity of thermal energy present. For example, a firework sparkler is at a much higher temperature than a red-hot poker, but because there is much less mass in a firework sparkler than a poker there is more heat in the poker – which is why one can hold a sparkler but would be reluctant to hold a glowing poker.

Ice lines Ice lines occur at the distance from a star where different molecules freeze from gas to a solid. The distance of these ice lines from the central star will depend on the temperature of the star. For a Sun-like star the water ice line is around 3 AU, followed by the CO_2 ice line (CO_2 ice is known as dry ice here on Earth), ammonia ice line, followed by the methane ice line.

Inflation An extension of the standard Big Bang theory that suggests that the Universe expanded at a greatly increased rate for a short period less than a second after the Big Bang. While direct evidence for inflation and a theoretical understanding of its causes has so far proved elusive, it provides an elegant solution to several observational problems with the standard Big Bang theory.

Ionization Energetic photons can knock electrons away from their atomic nuclei, which are then said to be ionized. In the energetic conditions immediately after the Big Bang, the electrons had too much energy to be captured by atomic nuclei and the entire Universe was ionized. As the Universe expanded, it cooled until the electrons could be captured by the nuclei in order to form neutral atoms. Upon the advent of the first sources of light, the electrons were liberated once more at an epoch known, rather confusingly, as reionization.

Isotope An atom of the same element with the same number of protons but a different number of neutrons. The amount of a particular isotope something has can be used to age substances. And the amount of one isotope of a substance compared to another isotope, such as hydrogen to deuterium can be used to determine the origins of a substance in this case where our oceans came from.

Light-year The distance travelled by light in a year when passing through a vacuum, equivalent to 9.5×10^{15} metres or nearly six thousand billion miles. The Sun is eight light-minutes away – the light we see left the star eight minutes before – the nearest other star is 4.2 light-years away. The Sun lies 26,000 light-years from the centre of the Milky Way Galaxy, which is itself 100,000 light-years across. Bodies located 13 billion light-years away are seen as they appeared just after the Big Bang.

Luminosity The luminosity of a light source reflects the rate of emission of light. In other words, the luminosity of a star reflects its intrinsic rather than apparent brightness. The Sun appears much brighter than the other stars in the sky because it is close to us, even though many stars are much more luminous than our own rather ordinary star.

Magnitude The traditional measure of brightness for astronomical objects. The scale is rather confusing; the lower the number, the brighter the source appears. By definition, the bright star Vega has a magnitude of 0.0 and a difference of five magnitudes corresponds to a difference of 100 times in brightness. Vega is therefore 100 times brighter than a star with a magnitude of 5. In dark skies, the naked eye can see to a magnitude of 6 or so. These are apparent magnitudes, but it is also common to refer to absolute magnitudes. These reflect the luminosity of the source and are defined as the apparent magnitude the source would have at the standard distance of ten parsecs.

Mass There are two scientific definitions of mass. The first is the property of a body to resist acceleration; it takes more effort to push a car than a football. The second is the property of a body that defines the strength of its gravitational attraction; objects with more mass have a stronger gravitational pull. The two turn out to be equivalent so that the same definition of mass can be used for both. A common mistake is to confuse mass with weight. Weight is the force exerted on an object by gravity. When Neil Armstrong stepped out onto the Moon's surface, his mass did not change but his weight certainly did.

Meteor A shooting star, or meteor, is caused by the entrance of a small particle, usually no bigger than a grain of sand, into the Earth's atmosphere. The friction of the atmosphere causes the particle to burn up, leaving a short-lived rapidly moving trail across the sky, which we see as a meteor. Many of these dust grains are associated with comets; as the comet makes repeated passes through the inner Solar System, dust thrown off from the nucleus spreads out along its orbit. When the Earth's orbit intersects that of the comet, we see a meteor shower. Meteors from the same shower will appear to trace back to a single area of sky, known as the radiant – the effect is similar to standing on a motorway bridge and seeing the two parallel carriageways appear to merge in the distance. Of particular note are the two most famous showers; the Perseids, which have their maximum in August, are the most reliable annual shower while the Leonids produce spectacular meteor storms on a roughly 33-year period. During such a meteor storm, for a short period of only an hour or so, the rate may approach a meteor per second.

Meteorite A large, usually asteroidal body, which has survived entry through the Earth's atmosphere and landed on the planet's surface. Remarkably, there is no record of anyone being injured by a meteorite fall, although several cars have been badly damaged in recent years! Although the source of most meteorites is the Asteroid Belt (or other asteroid-sized bodies), a few are likely to have come from the Moon or Mars. The most famous of these Martian meteorites, ALH84001, includes structures that look intriguingly like terrestrial bacterial fossils, but the scale is different. There is the possibility that they are there due to contamination following the meteorite's arrival on Earth, but they remain the most tangible evidence that life might exist, or has existed, on Mars. The majority of meteorites are found today in Antarctica, where they stand out against the icy background.

Milky Way A luminous band of faint stars that crosses the sky, containing many nebulae and dust clouds in addition to stars. It is the projection of the disk of our own Galaxy, which is also known as the Milky Way, onto the celestial sphere.

Nebula From the Latin word for 'mist', 'fog' or 'cloud', the term 'nebula' is used in astronomy to refer to any visible mass of gas and dust. The most famous nearby example, the Orion Nebula, is a region in which stars are forming from condensing gas and dust. The newly formed stars are then able to light up the surrounding gas in what is known as a 'reflection nebula'. In the latter stages of a Sun-like star's life it will expel its outer layers, forming a 'planetary nebula'. The name comes from the often disk-like appearance in small telescopes, but is somewhat unfortunate as there is no association with

planets or with reflection nebulae. Dark nebulae, composed of dust blocking out light from more distant sources, are also observed; the most famous of these is the Coal Sack in the southern constellation of Crux.

Neutrino Small, lightweight particles that are produced as by-products of the nuclear fusion that powers stars such as the Sun. For many years, it was believed that they might be massless, but it is now clear they have a certain amount of mass (although not enough to account for the required amount of dark matter). This in turn solved the long standing 'neutrino problem' in which the number of neutrinos observed from the Sun was much lower than expected by theory. The tiny mass allows the neutrinos to change between three 'flavours' – electron, mu and tau neutrinos – en route between the Sun and the Earth, and previous generations of detectors were sensitive only to the least massive flavour. The number of neutrino flavours can be predicted by the Big Bang theory, and provides an excellent test of the idea that the Universe began in a hot, dense state.

Neutron Neutrons are one of the two types of particles – both composed of three quarks – that make up atomic nuclei. They weigh almost the same as protons, but carry no electric charge. Under the extreme conditions of a supernova explosion, protons and electrons can combine to form neutrons, resulting in a dense neutron star being produced from the dying star's core. The maximum mass for a neutron star is believed to be around eight solar masses; any larger than this and collapse to a black hole is inevitable.

Nucleus The nucleus of an atom is made up of positively charged protons and neutral neutrons, and contains almost all of the mass of the atom. At the high temperatures and pressures in the centres of stars, electrons are too energetic to be captured by the positively charged nucleus; so it is atomic nuclei that combine in fusion to form heavier elements. The number of protons in the nucleus of an atom defines its type, so that hydrogen has a single proton, helium two, lithium three and so on.

Parsec A unit of distance equal to 3.26 light-years. Seen from a distance of 1 parsec, the Earth would appear in the sky just 1 arcsecond (or a 3600th of a degree) away from the Sun.

Planck time In quantum mechanics, this is the smallest possible 'unit' of time, equal to 5×10^{-44} seconds. Even if a clock was accurate enough to measure a smaller period, quantum mechanics makes that impossible. Whether this is a real feature of the Universe or a statement of the inadequacy of quantum mechanics remains to be seen.

Positron The antiparticle of the electron, a positron has the same mass as an electron but the opposite charge. Like all matter–antimatter pairs, an electron and positron will annihilate on collision producing only energy.

Proton A positively charged particle composed of three quarks; protons are one of the two components that make up atomic nuclei.

Protoplanetary disk A protoplanetary disk is the material surrounding a young protostar (before the star fully ignites becoming a main sequence star) that contains the gas and dust from the original solar nebula that has not become part of the star. The protoplanetary disk is where the planets will form.

Pulsar A rapidly spinning neutron star (produced in a collapsing supernova) will produce radiation in a thin beam from near both poles. As the star rotates, so this beam sweeps, lighthouse-like, across the sky. If it happens to cross the Earth, we see a rapidly pulsing source. So regular are these pulses that the first detection was labelled 'LGM-1', standing for Little Green Men-1! There is one known example of a double pulsar, and scientists are able to exploit the information from its pulses to provide stringent tests of the theory of general relativity.

Quantum The fundamental insight of quantum mechanics is that a particle cannot have any arbitrary amount of energy, but must have a whole number of small 'bricks' of energy. These building blocks are known as quanta ('quantum' is the singular form). In our everyday lives, the effects of this phenomenon are small, because a single quantum is an extremely small amount; on the scales familiar to atoms and molecules, however, things are very different.

Quark The particles that combine to form protons, neutrons and other, more exotic, particles, quarks are now believed to be fundamental particles. In other words, they cannot be split. They come in six 'flavours', whimsically named 'up', 'down', 'strange', 'charm', 'top' and 'bottom'. (There has been a recently unsuccessful movement to rename the last two 'truth' and 'beauty'.) Quarks are attracted to each other by the strong nuclear force, and have the remarkable property that they can never be isolated as single quarks. The strong force increases with distance, so if two quarks are pulled apart the force attracting them to each other actually increases!

Quasar The original definition of quasar, or quasi-stellar object, was a star-like source that appeared to be at a great distance. Decades of observations have revealed that they are in fact galaxies harbouring extremely massive black holes at their centres, which are in the process of consuming huge amounts of dust and gas. This in-falling material radiates as it falls toward the central black hole, and this powerful source of radiation is responsible for our ability to see quasars from the most distant parts of the Universe. More common in the distant past, it has recently been suggested that all galaxies may have experienced a 'quasar-like' phase, relaxing to become 'normal' galaxies only when the reservoir of material to feed the central black hole has been exhausted.

Redshift The movement of spectral features toward the red end of the spectrum of a receding source, due to the Doppler effect. Astronomers also use redshift as a co-ordinate of time; the present day is anequivalent to a redshift of 0, and the observed redshift increases as we look back toward the earliest stages of the Universe's evolution. The most distant source yet observed is at a redshift of 6.4, which is equivalent to just 870 million years after the Big Bang (and 12.9 billion light-years away).

Spectrum Electromagnetic radiation passed through a prism (or a fine grating) will split into its component wavelengths, an effect most familiar from the sight of a rainbow in the sky. This is known as a spectrum, and the relative intensities of different wavelengths can encode a huge amount of information about the object that emitted the light. In particular, a series of dark or bright lines known as spectral lines (see p.58) acts as a fingerprint for each of the elements present in the source, allowing astronomers to identify the composition of even the most distant objects. Sir Isaac Newton coined the word spectrum from the Latin for 'to see.'

Standard form Standard form is the name given to the scientific notation used for very small and very large numbers. It takes the form of a number between 1 and 10, and then a factor by which that number should be multiplied. Rather than writing one and a half million as 1,500,000 it would be represented as 1.5×10^6. The factor above 10 represents the number of zeros to be put to the right of the first number. Similarly, then, a billion is written as 1.0×10^9 and a millionth as 1.0×10^{-6}.

Steady-state theory A now discredited rival to the Big Bang theory, which held that the Universe was in a constant state of continued expansion and small-scale creation of matter.

Strong nuclear force The force that binds quarks together to form larger particles such as protons and neutrons. It increases with distance, so that as two quarks are pulled apart the force between them increases.

Supermassive A term usually applied to the black holes at the centre of galaxies. An exact definition is elusive, but it is usually taken to mean several million times the mass of the Sun.

Supernova When a large star has used up the fuel in its centre, it suddenly collapses. The increased pressure at the centre will lead to the formation of a dense remnant such as a black hole or neutron star, at which point the majority of the remaining material will rebound outward in a powerful explosion known as a supernova. Such an event can easily outshine all the other stars in a host galaxy for a period of a few weeks before gradually fading. Exceptions to this general picture are type 1a supernovae, which are produced in binary systems in which material from a large star is able to build up on the surface of a white dwarf before reigniting when a critical density is reached. Somewhat frustratingly, there has not been an observed supernova in the Milky Way since the invention of the telescope; the nearest was in the satellite galaxy the Large Magellanic Cloud in 1987.

Wavelength The distance between two crests of a wave. The wavelength of red visible light is 4.0×10^{-7}m, while radio waves can have wavelengths of many kilometres.

Weak nuclear force The force responsible for particular kinds of radioactive nuclear decay.

Wormhole A purely theoretical structure (so far) that would allow distant regions of space to be connected by a 'short cut'. It has been speculated that black holes might mark one end of such a pathway, with material that descends into them re-emerging in a 'white hole'.

INDEX (Numbers in **bold** indicate illustrations)

FURTHER READING

The Dark Side of the Universe, Iain Nicolson (Canopus/Johns Hopkins University Press, 2007)

Introducing Astronomy: A Guide to the Universe, Iain Nicolson (Dunedin, 2014)

Patrick Moore's Data Book of Astronomy, Patrick Moore and Robin Rees (Cambridge University Press, 2011)

The Crowd and the Cosmos, Chris Lintott (Oxford University Press, 2019)

The Cosmic Tourist, Brian May, Patrick Moore, Chris Lintott (Carlton, 2012)

Cosmic Clouds in 3-D: Where Stars are Born, David Eicher and Brian May (London Stereoscopic Company, 2020)

A Survey of Radial Velocities in the Zodiacal Dust Cloud, Brian May (Springer, 2008)

Stargazers: Copernicus, Galileo, the Telescope and the Church, Allan Chapman (Lion, 2014)

PICTURE CREDITS

Every effort has been made to acknowledge correctly and contact the source and/or copyright holder of each picture, and Canopus Publishing Limited apologises for any unintentional errors or omissions, which will be corrected in future editions of this book.

Preliminary pages
pp. 4–5: J-P Metsavainio; p. 8: Petr Horálek/ESO; p. 9 Patrick Moore; p. 11: J-P Metsavainio; pp. 12–13 Denis Pellerin.

Introduction
pp. 14–15: Nik Szymanek; p. 16: NASA; p. 17: NASA; p. 18: H J P Arnold; p. 19: NASA/ESA/Hubble/K. Noll; p. 20: European Southern Observatory; p. 21: NASA; p. 22: James Symonds.

Chapter 1
pp. 24–5: Brian Smallwood; p. 27: Greg Parker/Noel Carboni, New Forest Observatory; p. 28: M.C. Escher's "Cubic Space Division"© 2006 The M.C. Escher Company-Holland. All rights reserved; p. 29: European Southern Observatory; pp. 30–1: James Symonds; p. 32: Nanoscale Science Laboratory, Cambridge; p. 34: Courtesy of Brookhaven National Laboratory; p. 35: G.T. Jones, Birmingham University / Fermi National Accelerator Laboratory; p. 36: W. Purcell (NWU) et al., OSSE, Compton Observatory, NASA; p. 37: James Symonds; p. 38: NASA/ESA/G. Illingworth, D. Magee and P. Oesch, University of California, Santa Cruz, R. Bouwens, Leiden University; and the HUDF09 Team; p. 39: James Symonds; p. 40: James Symonds; p. 41: Brian May.

Chapter 2
pp. 42–3: Brian Smallwood; p. 44: James Symonds; p.45: ESA/Planck Team; p. 46: NASA/GSFC/ JPL-Caltech; p. 47: NASA/ESA/P. Oesch and P. van Dokkum (Yale)/G. Brammer (STScI)/G. Illingworth (UC Santa Cruz); p. 48: NASA; p. 49: Royal Astronomical Society; p. 50: Kamioka Observatory, ICRR (Institute for Cosmic Ray Research), The University of Tokyo NSF; p. 51: NSF p. 52: BOOMERANG Collaboration; p. 53: Max-Planck- Institute for Astrophysics; p. 54: 2dFGRS Team; p. 55: James Symonds; p. 56: James Symonds; p. 57: SOHO; p. 58 (top): James Symonds; pp. 60–61: Caltech/University of Massachusetts; p. 62: NASA; p. 63: Dr. Christopher Burrows, ESA/STScI and NASA; p. 64: Anglo-Australian Observatory/David Malin Images; p. 65: European Southern Observatory; p. 66: Gamma-Ray Astronomy Team/NASA; p. 67: NASA and The Hubble Heritage Team (STScI); p. 68: James Symonds; p. 69: Event Horizon Telescope Collaboration.

Chapter 3
pp. 70–1: Brian Smallwood; p. 72: NRAO/AUI; p. 73: NASA/ESA/G. Illingworth, D. Magee and P. Oesch, University of California, Santa Cruz; R. Bouwens, Leiden

University, and the HUDF09 Team; p. 74: E.J. Schreier (STScI), HST, and NASA; p. 75: COBE/DIRBE/Richard Sword; p. 76 (top): Axel Mellinger; p. 76 (bottom): Patrick Moore; p. 77 (top): ESO; p. 77 (bottom): James Symonds; p. 78: NASA and The Hubble Heritage Team (STScI/AURA); p. 79 (top): Todd Boroson/NOAO/AURA/NSF; p. 79 (bottom left): /ESA/HST; p. 79 (bottom right): Jason Ware/www.galaxyphoto.com; p. 80: Mark Westmoquette (University College London), Jay Gallagher (Universityof Wisconsin-Madison), Linda Smith (University College London), WIYN//NSF, NASA/ESA; p.81 NASA-ESA/STScI/AURA/JPL-Caltech; p. 82: James Symonds; p. 83: X-ray: NASA/CXC/CfA/M.Markevitch et al.; Optical: NASA/STScI; Magellan/U.Arizona/D.Clowe et al.; Lensing Map: NASA/STScI; ESO WFI; Magellan/U. Arizona/D.Clowe et al.; p. 84: NASA/ESA and R. Massey (California Institute of Technology); p. 85 NASA/ CXC/M. Weiss; p. 86: James Symonds; P. 87: (top) James Symonds, (bottom) © Bettmann/ CORBIS; p. 88: ESA/Hubble/NASA; p. 89: NASA, Andrew Fruchter and the ERO Team [Sylvia Baggett (STScI), Richard Hook (ST-ECF), Zoltan Levay (STScI)] (STScI).

Chapter 4
pp. 90–1: Brian Smallwood; p. 92: European Southern Observatory; p. 93: Courtesy of Rogelio Bernal Andreo/ Wikimedia Commons; p. 94 (left): NASA ESA and The Hubble Heritage Team (STScI/AURA), acknowledgement William Blair (Johns Hopkins University); p. 94: (right) NASA/JPL-Caltech/R. Hurt (SSC/Caltech); p. 95: NASA, ESA, STScI, J. Hester and P. Scowen (Arizona State University); p. 99: NASA/ESA/STScI p. 100 ESA/NASA/JPL/CalTech; p. 101: Adam Block, Mt Lemmon SkyCenter, University of Arizona;p. 102: James Symonds; p. 103: James Symonds; p. 106: ALMA; p. 107 LMA (ESO/NAOJ/NRAO), S. Andrews et al., NRAO/AUI/NSF, S. Dagnello; p. 108: James Symonds; p. 109: Kate Shemilt; p. 110: ESO/VLT; p. 113 James Symonds; p. 114 NASA; p. 115: (top) ISAS/JAXA/Akatsuki, (bottom) Hubble; pp. 116-17: NASA / JPL-Caltech / SwRI / MSSS / Kevin M. Gill; p. 118 Cassini/JPL-Caltech/Space Science Institute; CaNASA/JPL/University of Arizona; 119: Lawrence Sromovsky/University of Wisconsin-Madison/W.W. Keck; 120 NASA/JP[L/USGS; NASA/JHUAPL/SwRI; p. 122-5: Petr Horálek/ESO .

Chapter 5
pp. 126–127: Brian Smallwood; p. 129: NAS/ESA; p. 132-3: ESA; p. 134: W. R. Normark/Dudley Foster; p. 135: National Oceanographic and Atmospheric Administration; 136: Brian May and Jamie Cooper; 137: Jamie Cooper; 138: Jamie Symonds; 139: (top) public domain, (bottom) Brian May; p. 140: Virgil L. Sharpton, University of Alaska, Fairbanks; p. 141: NASA; p. 142: (top) Jamie Symonds, (bottom) NASA; p. 144 (top) Patrick Moore, (bottom) Cassini/NASA; 145 (top and bottom) ESA/NASA/JPL/ University of Arizona; p. 146: (top) NASA, (bottom) Jamie Symonds.

Chapter 6
p. 148-9: Brian Smallwood; p. 150: (left) Brian Smallwood, (right)Brian May/Patrick Moore; p. 151: (left) Landsat Pathfinder Project, (right) Klim Levene; p. 152: (top): Phil James (Univ. Toledo), Todd Clancy (Space Science Inst., Boulder, CO), Steve Lee (Univ. Colorado), and NASA; p. 152: (bottom): NASA; p. 153: public domain; p. 154: NA/ JPL/Caltech/Space Science Institute; p. 155 (top) M. Montarges et al./ESO, (bottom) Jamie Symonds; p. 156: ESA & Garrelt Mellema (Leiden University, the Netherlands); p. 157: The Hubble Heritage Team (AURA/ STScI/NASA); p. 158: (top) Bruce Balick; p. 159: (top) NASA, ESA; Hans Van Winckel (Catholic University of Leuven, Belgium), and Martin Cohen (University of California, Berkeley); p. 159: (bottom) Andrew Fruchter (STScI) et al., WFPC2, HST, NASA; p. 160: (top) NASA , (bottom): NASA/SAO/CXC; p. 161: NASA; p. 162: NASA/CXC/ SSC/J. Keohane et al.; p. 163: J. M. Cordes & S. Chatterjee; p. 164: image courtesy of NRAO/AUI and HST/ STScI; p. 165: Brad Whitmore (STScI) and NASA; pp. 166–7: NASA, H. Ford (JHU), G. Illingworth (UCSC/LO), M.Clampin (STScI), G. Hartig (STScI), the ACS Science Team, and ESA; p. 168: LIGO; p. 169: VIRGO; pp. 170-1: ESA.

Chapter 7
p. 172-3: Brian Smallwood; p. 175: ESO; p. 176: Werner Benger; p. 177: Canada-France-Hawaii Telescope/J.- C. Cuillandre/Coelum; p. 179: Brian Smallwood;

Epilogue
p. 180: NASA.

Practical Astronomy
p. 182: Greg Parker/Noel Carboni.
p. 183: Kate Shemilt; p. 184: John Fletcher; p. 185: Jamie Cooper; p. 186: Ian Sharpe; p. 187: (top): Damian Peach; p. 187: (bottom): Brian May; p. 188: Jamie Cooper; p. 189: Damian Peach; p. 190: (top) Ian Sharpe, (bottom) Patrick Moore; p. 191 Brian May; pp. 192–16: all star maps courtesy James Symonds; pp. 197-8: © Instituto de Astrofisica de Canarias.